Honda 400 & 550 Fours Owners Workshop Manual

by John Witcomb

Models covered
Honda CB400 F. 408cc. UK March 1975 to 1977
Honda CB400 F. 408cc. US November 1974 to 1977
Honda CB550. 544cc. US September 1973 to 1976
Honda CB550 F1. 544cc. UK December 1975 to 1977
Honda CB550 F. 544cc. US October 1974 to 1977
Honda CB550 K. 544cc. US September 1974 to 1977

ISBN 978 0 85696 262 2

© Haynes Group Limited 1995

ABCDE
FGHIJ
KLM
3

Printed in India   *(262-1U8)*

Haynes Group Limited
Sparkford, Yeovil,
Somerset BA22 7JJ, England

Haynes North America, Inc
2801 Townsgate Road,
Suite 340
Thousand Oaks, CA 91361

Acknowledgements

Grateful thanks are due to Honda (UK) Limited for the technical assistance given so freely whilst this manual was being prepared and for permission to reproduce their drawings.

Brian Horsfall and Martin Penny gave the necessary assistance with the overhaul, and devised methods of overcoming the lack of service tools. Les Brazier took the photographs and Jeff Clew edited the text. Advice about tyre fitting was kindly supplied by the Avon Rubber Company Limited and NGK Spark Plugs (UK) Limited provided the spark plug photographs.

We are also greatly indebted to Arthur Vincent, of Vincent and Jerrom Limited, Taunton, for technical advice based on his experience as a leading Honda agent.

About this manual

The author learnt his motorcycle mechanics by trial and error - possibly more by error. In presenting this manual it is hoped that these errors can be avoided by others. Only by supervising the work himself, under conditions similar to those in which the amateur mechanic works, can the author ensure that the text is a true and concise record of procedure. Thus in the photographs, the hands shown are those of the author.

The machine selected had covered an average mileage, so that any problems encountered would be typical of those facing the average owner.

Honda service tools were not used, as generally an alternative method of removing or replacing a part was possible.

These motorcycles are designed to the metric system, and it is far easier to work metric. Metric spanners **must** be used.

A torque wrench should be begged or borrowed for use where torque wrench settings are given. Some car accessory shops and tool hire companies will often loan one.

Always have all tools and replacement parts to hand before commencing work. Baking trays or similar containers are useful for putting small parts in. Replace nuts and washers on the studs they fitted where possible, this avoids loss. Unless otherwise mentioned, reassembly should be carried out in reverse order to dismantling.

Each of the six Chapters is divided into numbered sections.

Within the sections are numbered paragraphs. Cross-reference throughout this manual is quite straightforward and logical. When reference is made, 'See Section 6.10', it means Section 6, paragraph 10 in the same Chapter. If another Chapter were meant it would say, 'See Chapter 2, Section 6.10'.

All photographs are captioned with a section/paragraph number to which they refer, and are always relevant to the Chapter text adjacent.

Figure numbers (usually line illustrations) appear in numerical order, within a given Chapter. 'Fig. 1.1' therefore refers to the first figure in Chapter 1.

Left-hand and right-hand descriptions of the machine and the components refer to the left and right of a given machine when normally seated.

Motorcycle manufacturers continually make changes to specifications and recommendations, and these, when notified, are incorporated into our manuals at the earliest opportunity.

We take great pride in the accuracy of information given in this manual, but motorcycle manufacturers make alterations and design changes during the production run of a particular motorcycle of which they do not inform us. No liability can be accepted by the authors or publishers for loss, damage or injury caused by any errors in, or omissions from, the information given.

Contents

Introduction
to the Honda 400cc and 550cc fours

The present Honda empire, which started in a wooden shack in 1947, now occupies a vast modern factory.

The first motorcycle to be imported into the UK in the early 60's, the 250 cc twin 'Dream', was the thin edge of a wedge which has been the Japanese domination of the motorcycle industry. Strange it looked too, to western eyes, with pressed steel frame, and 'square' styling.

In 1959, Honda commenced road racing in Europe, at the IOM TT races. They came 'to learn, next year to race, maybe', but walked off, with the manufacturers team award. A few years after this derided start, they were to dominate all classes, with such riders as Mike Hailwood, Jim Redman, and the late Tom Phillis and Bob McIntyre, on four, five and six cylinder machines. Even the previously unbeaten Italian multis no longer had things their own way, and were hard put to continue racing under really competitive terms.

Honda withdrew finally from racing in 1967, when at the top of the tree, holding 18 riders' world championships in all classes. In fact, in 1966, they held all five solo championships.

After their withdrawal from racing, Honda commenced work on a roadgoing version of their in-line four, scaled up to 750 cc. Without doubt, it was designed to be the number one 'Superbike' a position it has occupied since its introduction in 1969. With the experience gained from two year's production of the CB 750, and a demand for a lightweight touring motorcycle, the 500 cc four was introduced. This new model, well engineered and pleasantly compact, struck a happy medium between a sports and touring machine. The 500 was bored to 550 in 1974, and is the ancestor of the present CB 550F.

The 350 cc four, introduced into the USA shortly after the 500, never caught on. It was not marketed in the UK. The new CB 400F is based upon this previous design. In the short time that the 400 has been available, it has become apparent this model will prove to be the most popular yet produced.

Modifications to the
Honda 400cc and 550cc fours

In the short time that the CB 400F four has been in production, no modifications have been made. But there are differences between USA and UK models; in the layout of the electrics, the steering lock and the prop stand retractor.

The 550 models are little more than an over-bored CB 500. but there are minor differences in the case of the 1975 CB 550F model. There is a four-into-one exhaust system, which

necessitated re-jetting the carburettors. The front fork damping is re-designed, and the clutch operating mechanism is slightly changed. Engine oil capacity is raised by a half pint.

During the time these machines have been in production, the international standard for metric bolts has been altered. This means that certain sizes of bolt are not interchangeable. See 'Ordering Spare Parts'.

Above Right-hand view of the 1976 Honda CB 400F

Left The four-into-one exhaust system of the Honda CB 400F model

Ordering spare parts

When ordering spare parts for any Honda, it is advisable to deal direct with an official Honda agent, who should be able to supply most items ex-stock. Parts cannot be obtained from Honda (UK) Limited direct; all orders must be routed via an approved agent, even if the parts required are not held in stock.

Always quote the engine and frame numbers in full, and colour when painted parts are required.

The frame number is located on the side of the steering head, and the engine number is stamped on the upper crankcase, immediately to the rear of the two right-hand cylinders.

Use only parts of genuine Honda manufacturer. Pattern parts are available, some of which originate from Japan, but in many instances they may have an adverse effect on performance and/or reliability.

Honda do not operate a 'service exchange' scheme.

Some of the more expendable parts such as spark plugs, bulbs tyres, oils and greases etc., can be obtained from accessory shops and motor factors, who have convenient opening hours, charge lower prices and can often be found not far from home. It is also possible to obtain parts on a Mail Order basis from a number of specialists who advertise regularly in the motor cycle magazines.

Due to changes in the international standards (ISO), some hexagon and thread sizes have been altered on models after 1975. New spanners are required for the new size hexagons, but this does not affect interchangeability with old size bolts or nuts.

There has been a change in the pitch of the thread (the distance from crest to crest of the thread) for four diameters of thread; 3 mm, 4 mm, 5 mm and 12 mm. This means that these sizes of bolts or screws will not screw into tapped holes having the old pitch. Interchangeability is affected and great care must be taken to use the bolt, screw or nut having the correct thread pitch, to avoid damaging the thread in mating parts.

To avoid confusion with bolts, screws or nuts to the old standard, the latest fasteners are marked. All diameters of hexagon head bolts to the latest standard have a number, indicating the strength, embossed on the top of the head. Although this appears on all hexagon head bolts, only 3 mm, 4 mm, 5 mm and 12 mm bolts are not interchangeable.

All other bolts or screws having altered pitch, have a hemispherical bump or dimple on the top or side of the head. All nuts having the new thread pitch, will have a hemispherical dimple on the top or side.

Frame number on steering head

Engine number on top of the gearbox

Routine maintenance

1 Routine maintenance is a continuous process that must start as soon as the machine is first used. Maintenance should be regarded as an insurance policy, to help keep the machine in peak condition, and to ensure long, trouble free service. It has the additional benefit of giving early warning of any faults that may develop, and is a regular safety check to the obvious advantage of both rider and machine.

2 Maintenance must be carried out at the specified mileages, or on a calendar basis if the bike is not used frequently, whichever falls soonest. Do not try to extend the period between maintenance tasks, this applies especially to oil changes. Failure to change engine oil at the recommended mileage can cause serious engine damage, or increased wear. Wear will cause a drop in oil pressure, which can cause increased wear and so on.

3 No special tools are required for normal maintenance tasks, those in the toolkit supplied with the machine should prove adequate. A tyre inflator, tyre pressure gauge, feeler gauges and pressure-type cable oiler are desirable additions. The tasks are described fully in their appropriate Chapters.

4 In addition to the routine schedule, certain items should be checked every time the bike is used. These are engine oil level, tyre pressures, brake operation, rear chain and lights.

5 Those items which are governed by statute (such as lights, speedometer and tyres etc), must also be checked every time the bike is used. These will depend on the country or state in which the bike is being ridden.

Monthly or every 1,000 km (500 miles)

Check, lubricate and adjust rear chain.
Check tyre pressures.

Check oil level using the dipstick

Top-up the engine oil if necessary

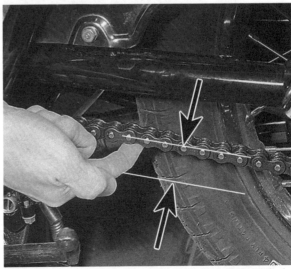

Check chain tension at the tightest point, in the middle of the bottom run

Check battery electrolyte level

Renew the oil filter

Check brake fluid level

Check the clearance from inner brake pad to disc

If the arrows on the rear brake wear indicator align, the linings require renewal

Grease the swinging arm pivot (CB 400F illustrated)

Check that lighting equipment works correctly.
Check the tightness of all nuts and bolts.

Three monthly or every 2,500 km (1,500 miles)

Change the engine oil when the engine is warm.
Check battery electrolyte level.

Six monthly, or every 5,000 km (3,000 miles)

Renew the oil filter.
Clean, adjust or replace as a set the spark plugs.
Clean, adjust or replace the contact breaker points.
Check the ignition timing.
Check the tappet clearance with the engine cold.
Adjust the cam chain.
Clean the air filter elements and the filter drain tube. (This should be done more often in dusty operating conditions).
Check and clean and adjust the carburettor, if necessary.
Examine, adjust and lubricate the throttle cables.
Clean the fuel filter.
Check the fuel pipes for cracks etc.

Check the clutch adjustment, and lubricate the cable.
Check the front brake fluid level.
Examine the brake shoes and pads for wear, and adjust the brakes if necessary.
Check the brake linkage and hoses for leaks or cracks. Lubricate the linkage pivots.
Check the wheels for true running, loose spokes etc.
Check the operation of the front and rear suspension.
Grease the swinging arm pivot and check for slackness.
Check prop stand operation (USA models only).

Yearly or every 10,000 km (6,000 miles)

Remove the sump and clean the oil strainer.
Renew the air filter element (this should be done more frequently in dusty operating conditions).
Drain and refill the front forks.
Adjust the steering head bearings, if necessary.
Check the front and rear wheel bearings.
Note that when, for example, doing the 5,000 km service, the 2,500 km and 1,000 km services must also be done.

Routine maintenance and capacities data

	CB 400F	CB 550	CB 550F
Fuel capacity	14 litres (3.1 Imp gall) (3.7 US gall)	14 litres (3.1 Imp gall) (3.7 US gall)	16 litres (3.5 Imp gall) (4.2 US gall)
Fuel tank reserve	3 litres (0.7 Imp gall) (0.8 US gall)	4 litres (0.9 Imp gall) (1.1 US gall)	3.7 litres (0.8 Imp gall) (1.0 US gall)
Engine oil capacity	3.5 litres (6.2 Imp pts) (3.7 US qts)	3 litres (2.6 Imp ats) (3.2 US qts)	3.2 litres (5.3 Imp pts) (3.4 US qts)
Engine oil grade	SAE 10W/40 or 20W/50		
Front forks oil capacity per leg (for refilling after draining)	145 cc (5.1 oz)	160 cc (5.4 oz)	160 cc (5.4 oz)
Fork oil grade	ATF or fork oil		
Spark plug type			
NGK	D-8ESL	D-7ES	D-7ES
ND	X-24ES	X-22ES	X-22ES
Spark plug gap	0.7 - 0.8 mm (0.028 - 0.032 in)	0.7 - 0.8 mm (0.028 - 0.032 in)	0.7 - 0.8 mm (0.028 - 0.032 in)
Contact breaker gap	0.3 - 0.4 mm (0.012 - 0.016 in)	0.3 - 0.4 mm (0.012 - 0.016 in)	0.3 - 0.4 mm (0.012 - 0.016 in)
Valve clearance (cold)			
Inlet	0.05 mm (0.002 in)	0.05 mm (0.002 in)	0.05 mm (0.002 in)
Exhaust	0.05 mm (0.002 in)	0.08 mm (0.003 in)	0.08 mm (0.003 in)
Front brake pad/disc clearance	0.15 mm (0.006 in)	0.15 mm (0.006 in)	0.15 mm (0.006 in)
Rear chain. Maximum total play ...	10 - 20 mm (7/16 - 7/8 in)	10 - 20 mm (7/16 - 7/8 in)	10 - 20 mm (7/16 - 7/8 in)
Tyre pressures (solo)			
Front	26 psi (1.8 kg/sq. cm)	26 psi (1.8 kg/sq. cm)	25 psi (1.7 kg/sq. cm)
Rear	28 psi (2.0 kg/sq. cm)	28 psi (2.0 kg/sq. cm)	28 psi (2.0 kg/sq. cm)
Tyre pressures (with passenger)			
Front	26 psi (1.8 kg/sq. cm)	26 psi (1.8 kg/sq. cm)	28 psi (2.0 kg/sq. cm)
Rear	36 psi (2.5 Kg/sq. cm)	36 psi (2.5 kg/sq. cm)	36 psi (2.5 kg/sq. cm)

Chapter 1 Engine, clutch and gearbox

Contents

Specifications

Engine				CB400F	CB550	CB550F
Type ...	...	...		Four cylinder, transverse, in-line. Air cooled		
Bore ...	...	...		51 mm (2.01 in.)	58.5 mm (2.3 in.)	58.5 mm (2.3 in.)
Stroke	...	...		50 mm (1.97 in.)	50.6 mm (1.99 in.)	50.6 mm (1.99 in.)

Compression ratio	9.4 : 1	9.0 : 1	9.1 : 1
Actual capacity	408 cc (24.9 cu in.)	544 cc (33.18 cu in.)	544 cc (33.18 cu in.)
BHP	37 at 8500 rpm (Din)	50.0 at 8500 rpm	50.0 at 8000 rpm
Engine rotation	Clockwise, viewed from right (alternator side)		
Valve operation	Single overhead camshaft, chain driven		
Valve timing:			
Inlet opens	5º BTDC	—	—
Inlet closes	35º ABDC	—	—
Exhaust opens	35º BBDC	—	—
Exhaust closes	5º ATDC	—	—
Valve clearance:			
Inlet	0.05 mm (0.002 in.)	0.05 mm (0.002 in.	0.05 mm (0.002 in.)
Exhaust	0.05 mm (0.002 in.)	0.08 mm (0.003 in.)	0.08 mm (0.003 in.)
Cylinder compression ...	12 kg/cm^2 (170.7 psi)	12 kg/cm^2 (170.7 psi)	12 kg/cm^2 (170.7 psi)
Idling speed	1,200 rpm	1,000 rpm	1,000 rpm
Engine oil capacity ...	3.5 litres (6.2 Imp. pints) (3.7 US quarts)	3.0 litres (2.6 Imp. quarts) (3.2 US quarts)	3.2 litres (5.3 Imp. pints) (3.4 US quarts)
Cylinder head flatness, max.	0.3 mm (0.0118 in.)	0.3 mm (·0118 in.)	0.3 mm (·0118 in.)
Gudgeon pin outside dia. min.	12.9 mm (0.5079 in.)	—	—
Small end inside dia. max.	13.10 mm (0.5158 in.)	—	—
Piston ring side clearance, max.	0.15 mm (0.0059 in.)	0.18 mm (0.007 in.)	0.18 mm (0.007 in.)
Piston ring gap,			
standard, top ...	0.15 - 0.35 mm (0.006 - 0.14 in.)	0.15 - 0.35 mm (0.006 - 0.014 in.	0.15 - 0.35 mm (0.006 - 0.014 in.)
second	0.15 - 0.35 mm (0.006 - ·014 in.)	0.15 - 0.35 mm (0.006 - 0.014 in.	0.15 - 0.35 mm (0.006 - 0.014 in.)
oil control	0.2 - 0.5 mm (0.008 - 0.20 in.)	0.3 - 0.9 mm (0.010 - 0.035 in.)	0.3 - 0.9 mm (0.010 - 0.035 in.)
Piston ring gap,			
max. top	0.7 mm (0.028 in.)	0.7 mm (0.028 in.)	0.7 mm (0.028 in.)
second	0.7 mm (0.028 in.)	0.7 mm (0.028 in.)	0.7 mm (0.028 in.)
oil control	0.9 mm (0.035 in.)	1.1 mm (0.043 in.)	1.1 mm (0.043 in.)
Crankshaft runout, max.	0.05 mm (0.002 in.)	0.05 mm (·002 in.)	0.05 mm (·002 in.)
Crankshaft journal clearance			
standard	0.018 - 0.048 mm (0.0007 - ·0019 in.)	0.020 - 0.046 mm (0.00079 - 0.00181 in.)	0.020 - 0.046 mm (0.00079 - 0.00181 in.)
max.	0.08 mm (0.0032 in.)	0.080 mm (0.0031 in.)	0.080 mm (0.0031 in.)
Big end side clearance, max.	0.15 mm (0.006 in.)	0.35 mm (0.0138 in.)	0.35 mm (0.0138 in.)
Big end journal clearance,			
standard	0.018 - 0.048 mm (0.0007 - ·0019 in.)	0.2 - 0.46 mm (0.00079 - 0.00181 in.)	0.2 - 0.46 mm (0.00079 - 0.00181 in.)
Big end journal clearance, max.	0.08 mm (0.0032 in.)	0.08 mm (0.0032 in.)	0.08 mm (0.0032 in.)
Primary chain guide thickness in centre, min.	5.0 mm (0.197 in.)	—	—
Camchain tensioner slipper thickness in centre, min	3.0 mm (0.118 in.)	—	—
Camchain guide thickness, min.	5.0 mm (0.197 in.)	—	—
Rocker arm to shaft clearance, max. ...	0.1 mm (·0039 in.)	—	—
Total height of cam, min.	28.0 mm (1.1024 in.)	35.85 mm (1.4075 in.) inlet 34.45 mm (1.3563 in.) exhaust	35.85 mm (1.4075 in.) inlet 34.45 mm (1.3563 in.) exhaust
Camshaft centre journal runout, max.	0.1 mm (·0039 in.)	0.1 mm (·004 in.)	0.1 mm (·004 in.)
Valve seat width	0.7 - 1.5 mm (·03 - ·06 in.)	1.0 - 1.5 mm (0.039 - 0.059 in.)	1.0 - 1.5 mm (0.039 - 0.059 in.)
Valve stem outside dia. min.	5.35 mm (·2106 in.)	—	—
Valve stem guide clearance, max.	0.3 mm (·0118 in.)	0.080 mm (0.0031 in.) inlet 0.010 mm (0.0039 in.) exhaust	0.080 mm (0.0031 in.) inlet 0.010 mm (0.0039 in.) exhaust
Valve spring free length, min.			
Inner	27.0 mm (1.063 in.)	34.5 mm (1.35 in.)	34.5 mm (1.35 in.)
Outer	32.5 mm (1.279 in.)	39.0 mm (1.53 in.)	39.0 mm (1.53 in.)
Gearbox	6-speed all indirect	5-speed all indirect	5-speed
Gear ratios 1 ...	2.733	2.353	2.353
2 ...	1.8	1.636	1.636
3 ...	1.375	1.269	1.269
4 ...	1.111	1.036	1.036

			400F	550	550F
5	...	0.965		0.900	0.900
6	...	0.866		—	—

	400F	550	550F
Primary transmission ...	HY - VO inverted tooth chain		
Selector fork finger width, min.	5.5 mm (0.216 in.)	—	—
Selector fork shaft dia. min.	12.9 mm (0.5079 in.)	—	—
Selector fork inside dia. max.	12.95 mm (0.5058 in.)	—	—
Transmission gear backlash, max.	0.2 mm (0.008 in.)	0.2 mm (·008 in.)	0.2 mm (·008 in.)
Transmission gear to shaft clearance, max.	0.2 mm (0.008 in.)	—	—
Clutch	Multi plate wet type		
Friction plate thickness, min.	2.3 mm (0.090 in.)	2.3 mm (0.090 in.)	2.3 mm (0.090 in.)
Clutch plate flatness, max. bow	0.1 - 0.2 mm (0.004 - 0.008 in.)	0.3 mm (0.011 in.)	0.3 mm (0.011 in.)
Clutch spring free length, min.	29.75 mm (1.1712 in.)	35.4 mm (1.39 in.)	35.4 mm (1.39 in.)
Clutch centre to clutch plate 'B' clearance	0.1 - 0.5 mm (0.004 - 0.020 in.)	—	—

Torque wrench settings	kg m	lb ft
Cylinder head nuts *	2.0 - 2.3	14.5 - 16.6
Crankcase bearing cap bolts - 8 mm	2.3 - 2.5	16.6 - 18.0
Connecting rod nut	2.0 - 2.2	14.5 - 15.9
Rocker cover bolt - 6 mm *	0.7 - 1.1	5.1 - 8.0
Cam sprocket bolt	1.8 - 2.0	10.8 - 12.2
Crankcase bolts - 6 mm - 400F only	0.7 - 1.1	5.1 - 8.0
Crankcase bolts - 550 models	2.3 - 2.5	16.6 - 18.0
Primary drive gear bolt - 400F only	3.0 - 4.0	21.7 - 29.0
Alternator rotor centre bolt - 550 models	4.0 - 4.2	28.9 - 30.3
Alternator rotor centre bolt - 400F	3.0 - 4.0	21.7 - 29.0
Clutch centre nut - 400F only	4.0 - 4.5	29.0 - 32.6
Exhaust pipe flange nut	0.8 - 1.2	5.7 - 8.6
Oil filter centre bolt	2.7 - 3.3	19.5 - 23.8

* *Tighten these down in several equal steps, to the final torque.*

1 General description

The in-line, four cylinder air-cooled, engine of the Honda CB 400F and CB 550 models is built in unit with the gearbox, and mounted transversely in the frame. The aluminium alloy die cast crankcase is separated horizontally, with all shafts, other than the gearbox input shaft, lying on the jointing line. Aluminium alloy is also used for the cylinder block, which has steel insert cylinder liners and for the cylinder head

The crankshaft is a one-piece forging, and runs in five journals with split shell bearings. The H-section connecting rods also have shell bearings in the split big-ends, whereas the small ends are plain bushes. The alternator is mounted directly on the left-hand end of the crankshaft and the contact breaker on the right-hand end.

The pistons have full skirts, and a flat top with cutaways for the valves. There are two compression rings and a three-piece oil control ring. The gudgeon pin is fully floating.

The single overhead camshaft, driven by an endless chain from the centre of the crankshaft, runs in plain unlined bearings in the cylinder head. The chain passes through a tunnel between the centre cylinders and is tensioned by a self-adjusting rubber-faced slipper. The rockers, which bear directly on the cams, are supported on shafts in the rocker cover.

Primary drive is by a short non-adjustable HY-VO inverted tooth chain, from the centre of the crankshaft. It drives the gearbox input shaft through a rubber vane type shock absorber. The right-hand end of the input shaft carries the straight cut spur drive pinion; the left-hand end drives the oil pump. The starter gear is also mounted on the input shaft, and drives it through a centrifugal clutch. The shaft is supported in two ball journal bearings. A spur gear meshes with the drive pinion, and transmits the drive to the countershaft via a multi-plate wet clutch.

The 400F model has a six speed gearbox and the two 550cc models five speeds. Gear selection is by means of dogs on sliding gears, which are moved by three selector forks. The forks are moved in turn by a rotary selector drum. The countershaft and mainshaft each have a ball journal bearing and a needle roller bearing.

A novel feature on the CB 400F is the external gearchange linkage, employing balljoints to transmit the motion of the foot change lever to the gearchange shaft lever. This permits better ergonomic positioning of the gear pedal.

2 Operations with engine/gearbox in frame

It is not necessary to remove the engine/gearbox unit from the frame unless the crankshaft or the gearbox requires attention. Theoretically, it is possible to give the engine unit a top overhaul with the engine in frame. Other components, such as the clutch, gear selector mechanism, contact breaker, alternator and oil pump are all readily accessible without need to remove the engine unit, although this may sometimes prove an advantage in terms of better access and more working space if several repair operations need to be carried out at the same time.

Note that the crankcase cannot be separated until the engine is removed from the frame and that access is not available to the gearbox until this has been accomplished. It is, however, possible to overhaul the gearbox, after separation, without need to disturb the cylinders. The crankcase must be rejoined before it can be replaced in the frame.

3 Operations with engine/gearbox removed

1 Removal and replacement of the main bearings and big-ends.
2 Removal and replacement of the crankshaft assembly.
3 Removal and replacement of the primary chain.
4 Removal and replacement of the gearbox components, selectors, kickstart mechanism (CB 550 models) and reduction gear.

4 Method of engine/gearbox removal

As described previously, the engine and gearbox are built in unit and it is necessary to remove the unit complete in order to gain access to either unit. Separation of the crankcase is achieved after the engine unit has been removed from the frame and refitting cannot take place until the engine/gear unit is assembled completely. Access to the gearbox is not available until the engine has been dismantled and vice versa in the case of attention to the bottom end of the engine. Fortunately, the task is made easy by arranging the crankcases to separate horizontally.

5 Engine and gearbox: removal from the frame

1 Put the bike firmly on its centre stand, preferably on a platform at working height.
2 Before commencing any work, disconnect the negative (earth) lead to the battery.
3 Remove the petrol tank. Turn off the petrol tap, and pull off the fuel pipes (including the petrol breather pipe at the junction under the rear of the tank, on the 550F models). Take off the tank rear retaining rubber before it falls off. It fits onto two ears on the frame.
4 Working will be found to be easier, if the seat is removed, although this is not essential (also, the seat won't get damaged). Take out the split cotter pins, and the seat hinge clevis pins. Remove the seat. Replace the pins to avoid loss.

5 Remove the air manifold and carburettors, as described in Chapter 3, Section 5. Pull the air manifold drain tube (when fitted) out of its locating hole at the bottom end. Remember where it went! Remove the carburettor assembly as a complete unit, with drain etc tubes. Lay it aside carefully in a clean place.
6 Pull off the spark plug leads, they are plainly numbered 1 - 4, left to right. Remove all spark plugs.
7 The rev-counter drive is adjacent to the right-hand rocker cap. Unscrew the crosshead screw, and extract the drive cable. Replace the screw.
8 Undo the six bolts holding the engine breather to the top of the rocker box. On the CB 400F model, the clutch cable is retained in a clip on the rear right-hand screw. On the CB 550 model, the outer two spark plug leads are in clips under the two outer screws. Take off the breather and gasket, pull the breather tube off of the filter housing (not on CB 550).
9 CB 400F model only: Disconnect the light grey with black, and black wires at the horn. Undo the horn mounting nuts; there is a wire cable clip under the front one. Remove the horn. Leave the bolts and cable harness bracket on the other side of the frame spine, replace the nuts.
10 Put a container, minimum capacity 4 litres (1 gallon) under the sump, remove the sump drain plug and drain all the oil. Also remove the oil filter after unscrewing the central bolt. Take care, because the filter housing is full of oil too. The oil will drain quicker if the engine is hot.
11 While the oil is draining, remove the frame cover on the left side. Pull back the solenoid cover, and undo the starter cable terminal nut. Remove the starter cable ring terminal, and replace the nut and washers. It may be easier to remove the flasher unit to get at the nut. The relay is held in a rubber clip.
12 Unplug the seven-pin plug, and the yellow and the blue wires from the contact breaker. These are in covers in front of the air filter and battery compartment respectively.
13 Remove the gearlever.
CB 400F model: Note the punch marks on the lever and splined shaft, to aid replacement in the same position. Lever off the gear pedal circlip and remove the washer. Remove the gearlever pinch bolt, and pull off the lever and pedal together. Replace the bolt.
CB 550 and CB 550F models: Remove the gearlever pinch bolt, and pull off the lever. Replace the bolt. Note the position of the lever for replacing.
14 Remove the starter motor cover and gasket on the CB 550 models, after unscrewing the two hexagonal screws. Remove the gearbox cover on the left-hand side. Note the position of the two locating dowels. There is no gasket. Remove the clutch cover on the right-hand side, there is no need to disturb the

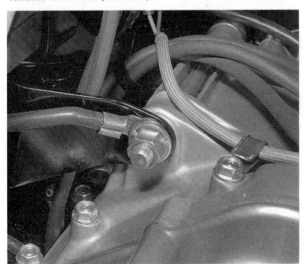

5.2 Remove the battery earth lead

5.7 Extract the rev counter drive cable

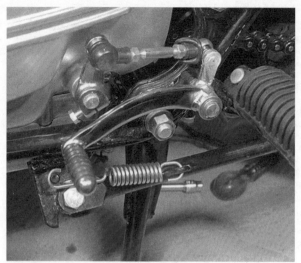

5.13 Pull off the CB 400F gear linkage as an assembly

5.15 Turn the sprocket retaining plate until it engages the splines

5.16 Remove the split collars and exhaust retaining rings

5.23a Remove the front engine plates (CB 400F shown) ...

5.23b ... and the lower front engine mounting bolts

5.24a There is a spacer between frame and crankcase

clutch mechanism. Leave the cover attached to the clutch cable, and hook it out of the way.

15 Before the chain is removed, and while the drive sprocket can be prevented from rotating, unscrew the sprocket fixing bolts. Take off the sprocket retaining plate by turning it until the splines can be engaged. Remove the sprocket.

16 Unscrew the eight bolts clamping the four exhaust retaining rings to the cylinder head.

17 CB 400F models: Unscrew the two bolts beneath the silencer, fixing it to the frame.
CB 550 models: Unscrew the bolts which retain each pair of silencers, and pillion footrests, to the hangers.
CB 550F model: Unscrew the bolt which retains the silencer to the pillion footrest hanger.

18 The exhaust system may now be taken off, complete with exhaust retaining rings and split collars. Hook up the chain.

19 If the oil has drained, take away the container. Remove the ten sump bolts (CB 550 models) or eleven sump bolts (CB 400F model), and drop the sump with gasket. Pull out the oil strainer (with spring on the CB 400F model).

20 Remove the nuts, with plain and spring washers on each end of the rear engine stud which secures the footrests. Remove the footrests.

21 Centre punch the brake pedal and shaft if not already marked to aid replacement in the same position. Remove the brake pedal pinch bolt and pull the pedal from the splined shaft. Replace the bolt.

22 Remove the nut on the right-hand end of the rear top engine mounting bolt and remove the battery earth lead terminal. Undo the two rear engine plate bolts and remove the plate on the CB 550 model, or swing it inwards on the CB 400F model. The lower nut is welded onto the plate on the CB 550 models and both nuts on the CB 400F model.

23 Remove the upper front engine mounting nut, bolt and washers. Also remove the engine plate to frame nuts, bolts and washers; followed by the engine plates. Remove the two lower front engine to frame bolts, nuts and washers. The nuts are trapped in the underside of the crankcase.

24 Support the engine and withdraw the upper rear engine bolt. There is a spacer between the frame and the crankcase on the left. Withdraw the lower rear engine stud; there is a spacer between the frame and the crankcase on the right, on CB 550 models. Lift the engine and remove it from the right; this is a two man lift. If you are not carefull, residual oil will drain out of the crankcase. If you can obtain two thick steel bars, which can be passed through the engine mounting lugs, manhandling the engine will be eased. Put the engine down on a clean flat surface.

6 Dismantling the engine/gearbox unit: general

1 Before commencing work on the engine, the external surfaces must be cleaned thoroughly. A motorcycle engine has very little protection from road dirt, which will sooner or later find its way into the dismantled engine if this simple precaution is not observed.

2 One of the proprietary engine cleaning compounds such as Gunk or Jizer can be used to good effect, especially if the compound is allowed to penetrate the film of oil and grease before it is washed away. When washing down, make sure that water cannot enter the carburettors or the electrical system, particularly if these parts are now more exposed.

3 Never use force to remove any stubborn part, unless mention is made of this requirement in the text. There is invariably good reason why a part is difficult to remove, often because the dismantling operation has been tackled in the wrong sequence.

4 Dismantling will be made easier if a simple engine stand is constructed that will correspond with the engine mounting points. This arrangement will permit the complete unit to be clamped rigidly to the workbench, leaving both hands free for the dismantling operation.

5 Have all spare parts, tools and containers for small items

(baking trays and egg cartons are useful) ready. An impact screwdriver will be required to loosen some screws - especially those retaining the crankcase covers. A torque wrench is essential - some motor accessory shops hire them.

6 Where differences between models occur, take care to follow the correct instructions.

7 Dismantling the engine/gear unit: removing the rocker gear and camshaft

1 This may be accomplished with the engine in the frame. Remove the breather cover after unscrewing the six bolts. Remove the rocker caps and slacken the tappets.

2 On the CB 550 models, remove the two rocker cover end covers after unscrewing the crosshead screws. Note the 'O' ring seals on the spigots.

3 CB 400F model: Slacken these fasteners uniformly, to relieve the stress gradually. Remove the rocker cover, noting the dowels at the rear. Undo ten screws and eight bolts - CB 550 models.

5.24b Remove the engine from the right

7.2 Remove rocker end covers (CB 550 models only)

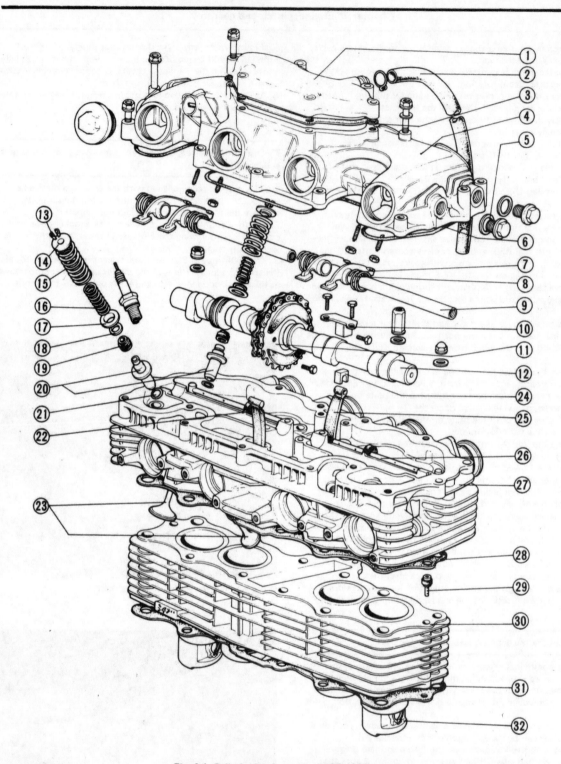

Fig. 1.1. Cylinder head and block, CB 400F model

1	Breather cover	9	Rocker arm shaft - 4 off	17	Outer seat - 8 off	25	Tensioner slipper
2	Breather tube	10	Cam chain tensioner holder	18	Inner seat - 8 off	26	Oil pipe - 2 off
3	Breather cover gasket	11	Cam sprocket	19	Valve stem seals - 4 off	27	Cylinder head
4	Rocker cover	12	Camshaft	20	Inlet and exhaust valve	28	Cylinder head gasket
5	Rocker spindle cap - 4 off	13	Valve cotter - 16 off		guide - 8 off	29	Oil restrictors - 2 off
6	Rocker cover gasket	14	Valve spring collar - 8 off	21	O-ring - 8 off	30	Cylinder
7	Valve rocker arm - 8 off	15	Outer valve spring - 8 off	22	Cam chain guide	31	Cylinder base gasket
8	Rocker arm side spring -	16	Inner valve spring - 8 off	23	Inlet and exhaust valves -	32	Oilway dowel - 2 off
	8 off				4 each of		
				24	Tensioner damper - 2 off		

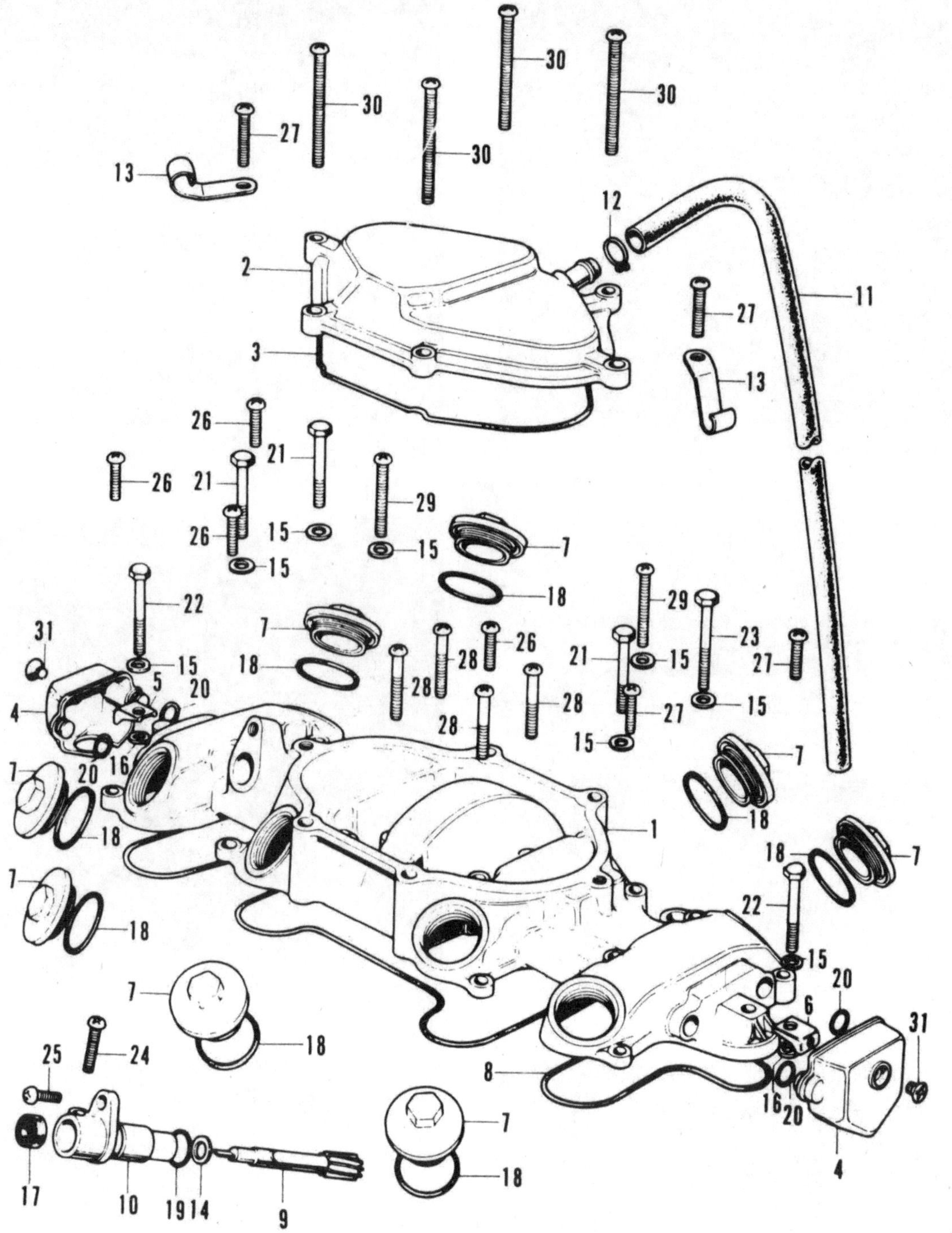

Fig. 1.2. Rocker gear, CB 550 models

1	Rocker cover	9	Tachometer gear	16	Aluminium washer - 2 off	24 Screw
2	Breather cover	10	Tachometer cap	17	Oil seal	25 Screw
3	Gasket	11	Breather tube	18	O-ring seal	26 Screw - 4 off
4	Rocker end cover - 2 off	12	Breather tube clip	19	O-ring seal	27 Screw - 4 off
5	End cover bracket, right	13	High tension lead clip - 2 off	20	O-ring seal	28 Screw - 4 off
6	End cover bracket, left	14	Thrust washer	21	Bolt - 3 off	29 Screw - 2 off
7	Tappet adjusting cap - 8 off	15	Washer - 8 off	22	Bolt - 2 off	30 Screw - 4 off
8	Cylinder head cover gasket			23	Bolt	31 Rocker end cover screw

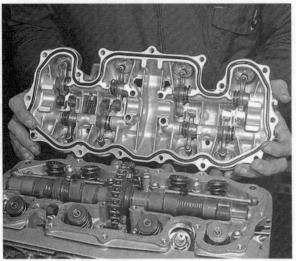

7.3 Remove the rocker cover

7.4a Pull out the oil jet tubes ...

7.4b ... they have different diameter spigots (CB 400F only)

7.5 Remove the cam chain tensioner holder (CB 400F model)

7.6 Slide the cam chain sprocket to the left

7.8 Unscrew the cam chain tensioner bolt (CB 550 models only)

Note the end cover brackets, with spacer washers, on the outer bolts.

4   CB 400F model only: Pull out the oil jet tubes behind the camshaft at each end. Each has a different diameter spigot to ensure correct reassembly. Note the grommet which supports the free end of the jet tube.

5   CB 400F model: Unscrew the two hexagonal screws securing the cam chain tensioner holder, this is to the rear of the cam chain tunnel. The screws are under spring tension. Remove the tensioner holder and tensioner. **Do not** slacken the chain tension adjuster at the bottom front of the cylinder block. CB 550 models: Loosen the cam chain adjuster locknut at the bottom rear of the cylinder block and turn the slotted adjuster fully clockwise. Tighten the locknut. This takes tension off of the cam chain.

6   Remove the contact breaker cover on the right, retained by two screws. Take care of the cork gasket. Turn the engine using a spanner on the large hexagonal washer, until one of the cam sprocket bolts comes to the top. Remove this bolt. Rotate the crankshaft 180°, until the second bolt can be removed. Slide the sprocket to the left, off of its locating shoulder, to obtain enough slack to unhook the chain (also to the left).

7   Hook a stout piece of wire, or a screwdriver, under the chain, to prevent it dropping into the tunnel while the camshaft and sprocket are removed from the right. Take the sprocket off of the camshaft.

8   CB 550 models only: Remove the cam chain tensioner bolt at the back of the cylinder head, above the adjuster.

8   Dismantling the engine/gearbox unit: removing the cylinder head

1   This may be accomplished with the engine in the frame. Remove the carburettors, see Chapter 3, Section 5. Remove the rocker cover as described previously.

2   CB 400F models: Remove the cam chain tensioner at the rear of the chain tunnel. Unhook and remove the chain guide at the front of the tunnel.
CB 550 models: Remove the six sealing rubbers from the inner nuts.

3   The cylinder head is clamped by 12 nuts, with an additional two bolts on the CB 550 models. Two of the nuts are close to the spark plug holes, take care not to drop them in! Take note of

8.3a Two nuts are close to spark plug holes

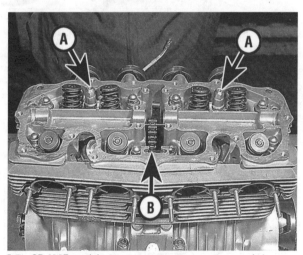

8.3b CB 400F models: Note the pillar nuts at the rear, (A) and cam chain (B)

8.3c These washers have bonded rubber inserts

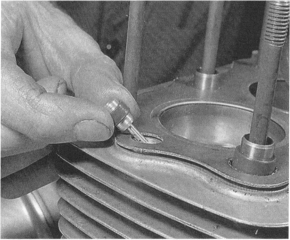

8.5 Note the dowel and oil restrictor, with gaskets (CB 400F)

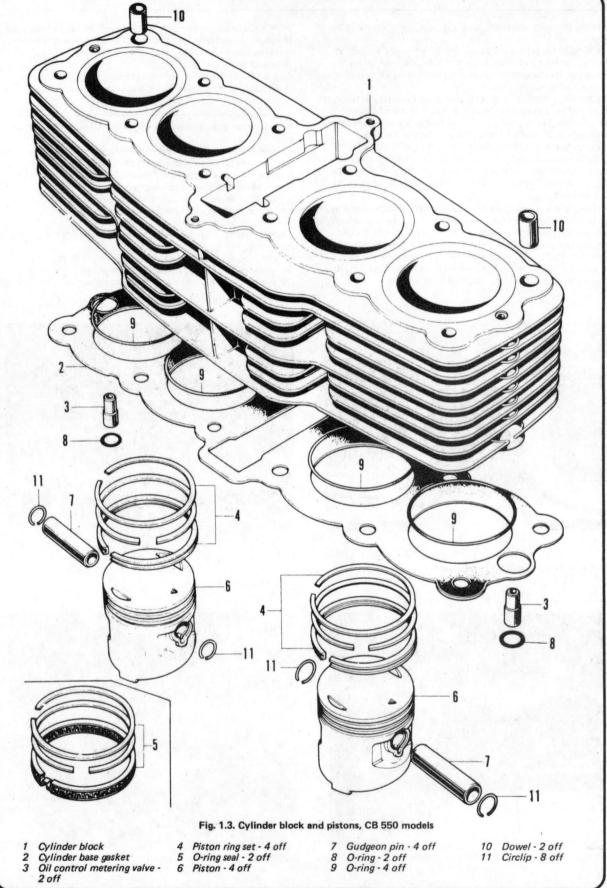

Fig. 1.3. Cylinder block and pistons, CB 550 models

1	Cylinder block	4	Piston ring set - 4 off	7	Gudgeon pin - 4 off	10	Dowel - 2 off
2	Cylinder base gasket	5	O-ring seal - 2 off	8	O-ring - 2 off	11	Circlip - 8 off
3	Oil control metering valve - 2 off	6	Piston - 4 off	9	O-ring - 4 off		

the positions of the nuts on the CB 400F model, two of them serve as pillars for fixing the rocker cover. Loosen the bolts (and nuts) in reverse of the order shown for tightening, to release stresses evenly. Remove the nuts and their copper or bonded rubber/metal washers. Note the positions of the two types of washer on the CB 400F model.

4   Lift off the cylinder head, allowing the chain to drop down its tunnel on the end of a wire hook. Put the head down on a clean flat surface. When the cylinder head has been removed, replace the cam chain support on the top of the cylinder block.

5   CB 400F model: Note the two dowels at the rear of the cylinder block, and the two dowels with tubular gaskets on the front outer studs. Also the two oil restrictors with tubular gaskets on the front outer studs. Also the two oil restrictors with tubular gaskets at each end, put these in a safe place.
CB 550 models: Note the two dowels at the rear of the cylinder block, and the two 'O' ring seals at each end. Put these in a safe place.

9   Dismantling the engine/gearbox unit: removing the cylinder block

1   Remove the cam chain guide from the cylinder block by raising the guide slightly, rotating it 90° and withdrawing it. Take care not to drop the cam chain into the crankcase during this operation. Unscrew the cam chain adjuster locknut and remove the tensioner unit from the cylinder block by pulling the unit upwards. If it is found difficult to remove the unit it can be easily released from the cylinder block after the block has been removed (CB 550 models only).

2   Pull up the cam chain and turn the crankshaft until all pistons are at the same level.

3   Lift the cylinder block, allowing the cam chain to drop down its tunnel. If the block is reluctant to come off, use a screwdriver between the joint faces (this is recommended by Honda!). Do not bruise the joint faces. If the joint is very tight try penetrating

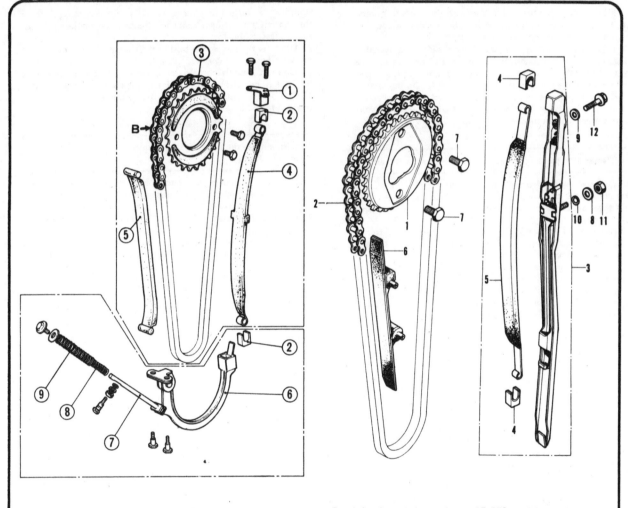

Fig. 1.4a. Cam chain tensioner, CB 400F model

1 Cam chain tensioner holder	*8 Tensioner outer spring*
2 Tensioner dampers - 2 off	*9 Tensioner outer spring*
3 Cam chain	*10 Plug*
4 Tensioner slipper	*11 Cam sprocket bolt - 2 off*
5 Cam chain guide	*12 Tensioner arm bolt - 2 off*
6 Cam chain tensioner arm	*13 Tensioner adjuster screw*
7 Chain tensioner adjuster	

Fig. 1.4b. Cam chain tensioner, CB 550 models

1 Camshaft sprocket	*7 Cam sprocket bolt - 2 off*
2 Cam chain	*8 Sealing washer*
3 Cam chain tensioner	*9 Aluminium washer*
4 Tensioner damper - 2 off	*10 O-ring seal*
5 Tensioner slipper	*11 Nut*
6 Cam chain guide	*12 Bolt*

oil down the stud holes and tap gently on the boss on which the capacity is cast.

4   Support the pistons as they emerge from the cylinder bores.

5   CB 400F model: Note there is cylinder locating dowels only. CB 550 models: Note the cylinder locating dowels, with 'O' ring seals, which double as oil metering jets.

10 Dismantling the engine/gearbox unit: removing the pistons and piston rings

1   Seal the crankcase mouth with clean rag, to prevent small items and dirt dropping in.

2   Remove the wire circlip from the inner end of each gudgeon pin, using a small screwdriver.

3   Push out each gudgeon pin; if it is tight warm the piston. This can be accomplished by wrapping it in hot wet rags, or by putting a hot flat iron on the crown. If the pin is knocked out, the connecting rod may be bent.

4   Mark each piston inside the skirt with the cylinder number (1 - 4, left to right), as it is removed. The crown is already marked to indicate the front.

5   Expand each piston ring carefully and slide it up over the ring lands. Do not expand the ring more than necessary, since it is very brittle and will break easily. Three thin strips of metal, slid between the ring and the piston, will help. On the later models a three-piece oil control ring is fitted. Remove each piece individually. Keep the rings with their respective pistons, so that they are not inadvertently mixed up.

11 Dismantling the engine/gearbox unit: removing the alternator

1   This is easily carried out with the engine in the frame. Disconnect the alternator wires (green, white and three yellow) alongside the oil pump, after removing the gearbox cover on the left-hand side.

2   Remove the alternator cover on the left-hand side, after unscrewing the four crosshead screws. Be careful with the gasket, and note the dowels.

3   Lock the crankshaft with a stout bar through the left-hand connecting rod eye (if the block is off). Slacken the rotor centre bolt.

4   Using a puller, bearing on the slackened off rotor bolt, extract the rotor. It may be broken from its taper by a blow with a soft

H.5025

Fig. 1.5. Method of removing and replacing piston rings

hammer when the puller is screwed up tight. Remove the rotor centre bolt and rotor.

12 Dismantling the engine/gearbox unit: removing starter motor

1   Remove the starter motor cover (CB 550 models only) and the gearbox cover on the left-hand side, (this may be carried out with the engine in the frame).

2   Unscrew the two bolts securing the starter motor to the rear face of the crankcase. Pull the motor to the left and remove. It may be quite tight in its housing.

13 Dismantling the engine/gearbox unit: removing the contact breaker and automatic advance unit

1   This may be accomplished with the engine in the frame. Remove the contact breaker cover. Mark with a centre punch the contact breaker back plate and the crankcase adjacent to it, to aid reassembly.

2   Unscrew the hexagonal bolt in the centre, and remove it with the large hexagonal washer.

3   Unscrew the three crosshead screws around the periphery of the back plate, and remove it. Unclip the contact breaker leads.

4   Pull off the cam together with the automatic advance unit; this is retained by a dowel only.

14 Dismantling the engine/gearbox unit: removing the clutch cover and clutch

1   If the clutch is being overhauled with the engine in the frame, first remove the right-hand footrest and brake pedal. Drain the oil from the crankcase.

2   Remove the kickstart pinch bolt and pull off the kickstart. Replace the bolt.

3   CB 400F model: Remove the triangular clutch adjuster cover, after unscrewing the two screws. Lever up the operating lever, and detach the cable nipple. It may be necessary to open up the slot in the shackle. Pull the nipple through the cable adjuster. CB 550 models: Unhook the clutch cable nipple from the external linkage at the clutch end, and pull the nipple through the adjuster.

4   Unscrew the ten crosshead screws and remove the clutch cover. Note the one clip on the lowest front screw of the CB 400F model or the three cable clips on the CB 550 models.

5   Undo the four bolts on the clutch spider evenly, to relax the springs. Take off the springs and clutch centre.

6   CB 400F models: Flatten the tab washer, and unscrew the clutch centre castellated nut. If the proper tool is not available, it is possible to do this using a thick piece of wire. Bend the wires to a 'U' shape, the right size to engage in opposite castellations on the nut. Find a socket spanner sufficiently large to jam over the wire, and unscrew the nut. The nut may be spoiled by this method, and a new one should be fitted. Remove the tab washer and the dished washer. CB 550 models: Extract the external circlip from the clutch shaft, followed by the shim.

7   Pull the clutch centre, plates, pressure plate and drum together from the shaft, followed by the thrust washer which is behind the drum.

15 Dismantling the engine/gearbox unit: removing the gearchange linkage

1   The gearchange linkage is behind the clutch and can be detached after removing the clutch. This may be carried out with the engine in the frame.

2   Remove the gear pedal.

10.2 Extract the gudgeon pin circlip

11.4 Use a puller to extract the rotor

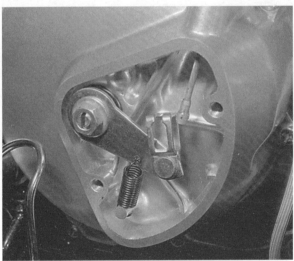

14.3 Detach the clutch cable (CB 400F)

14.5 Undo the four bolts on clutch spider (CB 400F)

14.6a Unscrew the clutch centre castellated nut ...

14.6b ... then remove the tab washer and dished washer (CB 400F)

3 CB 400F model: Unscrew the central bolt and remove with the large dished washer. Pull off the gear selector cam; this is dowelled to the selector drum. Take out the dowel if it remains in the selector drum, to avoid loss. Unhook the forked arm, and pull it out complete with gearlever shaft. Unscrew the pivot bolt, and remove the positive stopper (the 'Y' shaped lever, 8). Unhook the spring, unscrew the pivot bolt and remove the selector drum indexing lever (4). Unhook the spring, unscrew the pivot bolt and remove the neutral indexing lever (9). Note the position of all these parts and the location of their springs, for reassembly. The numbers in brackets refer to part numbers in Fig. 1.8.

4 CB 550 models: Unscrew the pivot bolt for the selector drum indexing lever (11), and the fixing bolt for the neutral indexing lever arm (15). The latter is just above the primary drive pinion and also secures the primary shaft bearing retaining plate.

Depress the gearchange lever and withdraw it with the gear pedal shaft. Note the location of all these parts, and the position of their springs for reassembly. If necessary, remove the selector drum cam after unscrewing the fixing bolt in the centre. If the cam locating dowel remains in the drum, remove it before it falls out. The numbers in brackets refer to Fig. 1.9.

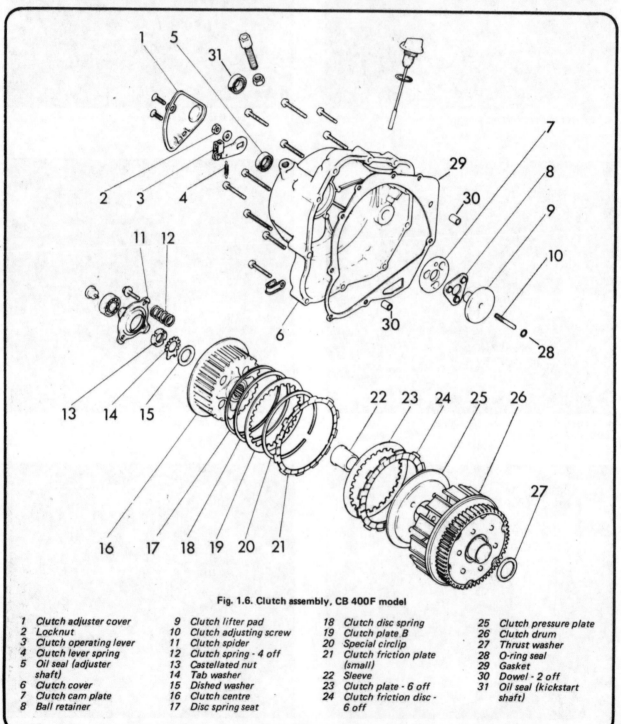

Fig. 1.6. Clutch assembly, CB 400F model

1 Clutch adjuster cover	9 Clutch lifter pad	18 Clutch disc spring	25 Clutch pressure plate
2 Locknut	10 Clutch adjusting screw	19 Clutch plate, B	26 Clutch drum
3 Clutch operating lever	11 Clutch spider	20 Special circlip	27 Thrust washer
4 Clutch lever spring	12 Clutch spring - 4 off	21 Clutch friction plate	28 O-ring seal
5 Oil seal (adjuster	13 Castellated nut	(small)	29 Gasket
shaft)	14 Tab washer	22 Sleeve	30 Dowel - 2 off
6 Clutch cover	15 Dished washer	23 Clutch plate - 6 off	31 Oil seal (kickstart
7 Clutch cam plate	16 Clutch centre	24 Clutch friction disc -	shaft)
8 Ball retainer	17 Disc spring seat	6 off	

14.7a Pull out the clutch centre ...

14.7b ... pressure plate and clutch drum

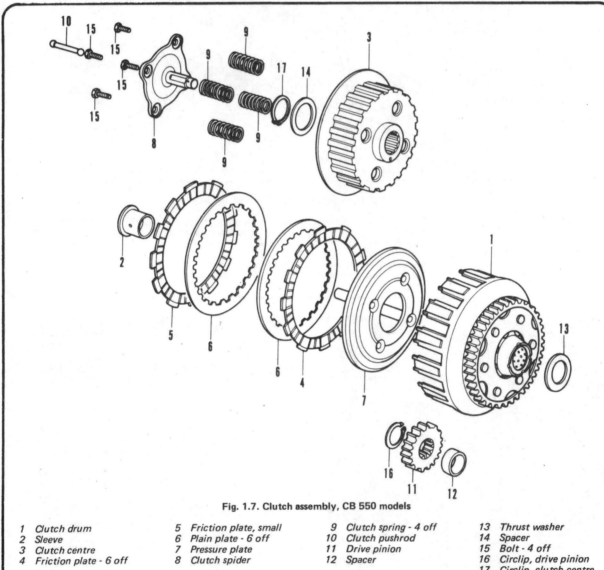

Fig. 1.7. Clutch assembly, CB 550 models

1	Clutch drum	5	Friction plate, small	9	Clutch spring - 4 off	13	Thrust washer
2	Sleeve	6	Plain plate - 6 off	10	Clutch pushrod	14	Spacer
3	Clutch centre	7	Pressure plate	11	Drive pinion	15	Bolt - 4 off
4	Friction plate - 6 off	8	Clutch spider	12	Spacer	16	Circlip, drive pinion
						17	Circlip, clutch centre

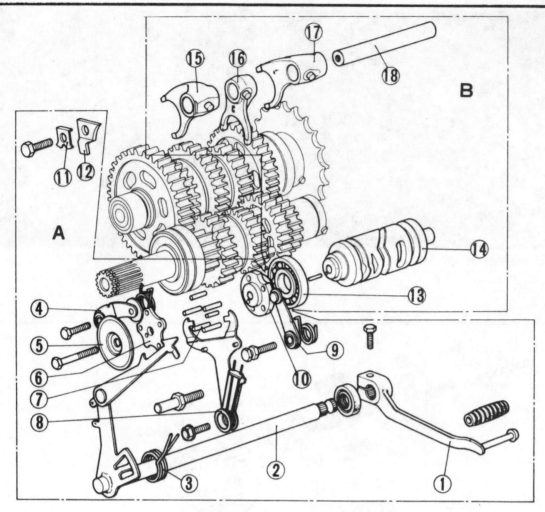

Fig. 1.8. Gearchange linkage, CB 400F model

1	Gearchange pedal	6	Selector drum outer cam
2	Gearshift spindle	7	Dowel - 6 off
3	Return spring	8	Positive stopper
4	Selector drum indexing lever	9	Neutral indexing lever
5	Dished washer	10	Selector drum inner cam

11	Tab washer
12	Bearing retaining plate
13	Ball journal bearing
14	Selector drum
15	Selector fork, right

16	Selector fork, centre
17	Selector fork, left
18	Selector fork shaft
19	Pivot bolt
20	Spring post
21	Pivot bolt
22	Retaining plate bolt

15.3a Remove the selector indexing lever ...

15.3b ... and neutral indexing lever (CB 400F)

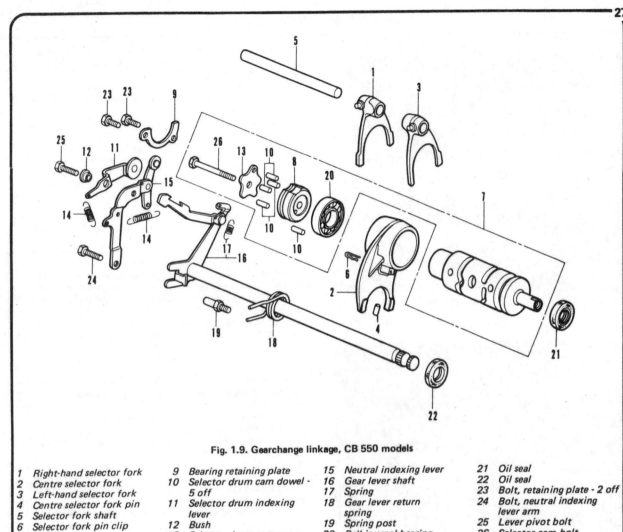

Fig. 1.9. Gearchange linkage, CB 550 models

1	Right-hand selector fork	9	Bearing retaining plate	15	Neutral indexing lever	21 Oil seal

1   Right-hand selector fork
2   Centre selector fork
3   Left-hand selector fork
4   Centre selector fork pin
5   Selector fork shaft
6   Selector fork pin clip
7   Selector drum
8   Selector drum inner cam

9   Bearing retaining plate
10 Selector drum cam dowel - 5 off
11 Selector drum indexing lever
12 Bush
13 Selector drum outer cam
14 Spring - 2 off

15 Neutral indexing lever
16 Gear lever shaft
17 Spring
18 Gear lever return spring
19 Spring post
20 Ball journal bearing

21 Oil seal
22 Oil seal
23 Bolt, retaining plate - 2 off
24 Bolt, neutral indexing lever arm
25 Lever pivot bolt
26 Selector cam bolt

15.4a Remove the selector indexing lever ...

15.4b ... and neutral indexing arm (CB 550 models)

16.2 Withdraw the kickstart shaft (CB 400F)

17.2a Unscrew the four bolts (CB 400F) ...

17.2b ... and remove the oil pump

16 Dismantling the engine/gearbox unit: removing the kickstart return spring and shaft

1   Remove the clutch cover. There is no need to disengage the clutch cable if engine is in frame.
2   CB 400F models: Unhook the return spring and withdraw the kickstart shaft.
3   CB 550 models: Only the spring can be removed; the shaft has to be removed after separating the crankcase. Remove the washer from the end of the shaft, then extract the spring retaining circlip. Unhook the spring from the crankcase and withdraw it.

17 Dismantling the engine/gearbox unit: removing the oil pump

1   Disconnect the oil pressure switch wire.
2   CB 400F models: Unscrew the one large and three small hexagonal bolts and remove the pump. There is a cable clip under the uppermost bolt. Note the two large and two small dowels with 'O' ring seals.
3   CB 550 models: Unscrew the three crosshead screws which secure the pump to the crankcase and remove the pump. Note the two dowels with 'O' ring seals.

18 Dismantling the engine/gearbox unit: separating the crankcase

1   The crankcase must be separated to obtain access to the crankshaft transmission; and in the case of the CB 550 models, to the kickstart mechanism.
2   Remove the wire from the neutral indicator switch. Pull the wire from the switch on the CB 550 models. Push down the spring loaded washer, and pull out the wire on the CB 400F model. Remove the wiring harness.
3   CB 400F models: Remove the primary drive pinion and dished washer after unscrewing the centre bolt. Pull out the shaft from the left (oil pump) side. Remove the plain sleeve from the drive side. Extract the bearing internal circlip, followed by the ball journal bearing and the flanged sleeve.
CB 550 models: Remove the primary shaft bearing retaining plate after unscrewing the one remaining hexagonal bolt. Find a long bolt that will screw into the tapped end of the primary shaft and a heavy weight (such as a socket spanner) that will slide on the shaft. Screw the bolt into the shaft, and extract the shaft by striking the weight against the head of the bolt. The shaft is tight and the drive side bearing will come out with it. The starter gear thrust washer should also come out with the shaft. (A Honda service tool, 07009-32301 is available for withdrawing the shaft).
4   CB 400F models: Remove the three 8 mm bolts, and the twelve 6 mm bolts in the top of the crankcase. Note the position of two cable clips. Invert the crankcase and remove the two 12 mm bolts at the front and the ten 12 mm bearing cap bolts. Slacken these gradually, in reverse of the order shown in Fig. 1.20a.
CB 550 models: Remove the three 6 mm and three 8 mm bolts from the top of the gearbox casing. Invert the crankcase and remove the ten 8 mm bolts and twelve 6 mm bolts. Note position of the cable clip. Slacken the 8 mm bolt gradually, in reverse of the order shown in Fig. 1.20b. Note the 6 mm bolts inside the sump and the three at the rear of the gearbox casing.
5   The crankcase may now be separated. If the engine is left in this inverted position, all the shafts will remain in the top half. Tap the crankcase gently with a soft hammer around the joint, then lift off the lower half. Do **not** lever the joint faces with a screwdriver.
6   **Note:** If the cylinder block has been removed, take care not to damage the cylinder studs when the engine is upside down.

18.3a Withdraw the primary shaft (CB 550 models)

18.3b Extract the primary shaft from oil pump side (CB 400F)

18.3c Remove the circlip and ball journal bearing

18.5 Don't forget this bolt inside the sump on CB 550 models

19.3a Remove the chain tensioner arm (CB 400F only)

19.3b Note O-ring seal on tensioner adjuster

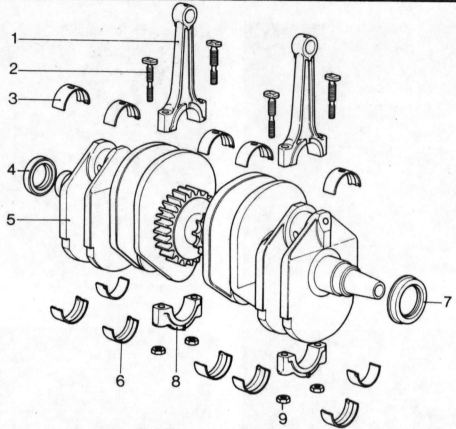

Fig. 1.10. Crankshaft and connecting rods

1	Connecting rod - 4 off	3	Crankshaft bearing - 10 off	6	Connecting rod bearing - 8 off	8	Connecting rod cap - 4 off
2	Connecting rod bolt - 8 off	4	Oil seal	7	Oil seal	9	Connecting rod nuts - 8 off
		5	Crankshaft				

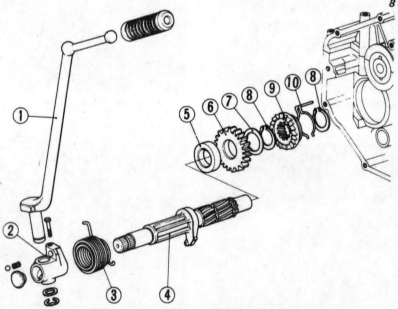

Fig. 1.11a. Kickstart assembly, CB 400F model

1	Kickstart lever	6	Kickstart pinion
2	Kickstart pivot	7	Thrust washer
3	Kickstart return spring	8	Circlip - 2 off
4	Kickstart shaft	9	Kickstart ratchet
5	Spacer	10	Starter pinion friction spring

19 Dismantling the engine/gearbox unit: removing the crankshaft

1   Having taken off the lower crankcase half, all the internals will be retained in the top half.
2   Lift the primary shaft drive sprocket and unhook the primary chain. Be careful when lifting the sprocket clutch rollers, or starter gear needle roller bearing. Both these items may fall out.
3   CB 400F models only: Remove the chain tensioner arm after unscrewing the two bolts at the front of the crankcase. Remove the tensioner adjuster and spring by slackening the tensioner adjuster bolt. It is not necessary to remove this bolt - it has an O-ring seal.
4   Lift out the crankshaft, it may be stubborn, and unhook the two chains. Remove the oil seals from each end.
5   The main bearing shells will remain in the crankcase. Do not remove them unless they are to be renewed.

20 Dismantling the engine/gearbox unit: removing connecting rods

1   Unscrew the big-end bearing cap nuts and pull off each cap. Remove the connecting rods. Mark each cap and rod to enable replacement on the same journal.
2   Replace the cap on the connecting rod, do not remove the bearing shells unless they are to be renewed.

21 Dismantling the engine/gearbox unit: removing the gearshafts and gear selectors

1   It is not necessary to remove the cylinder heads and block to work on the transmission.
2   Having removed the lower crankcase half, unhook the primary chain from the primary drive sprocket. Take care that the starting clutch rollers or the starter gear needle bearing do not drop out of the sprocket. Lift out the mainshaft and countershaft, each complete with gears and bearings. Remove the mainshaft oil seal from the crankcase if it has not come out with the shaft.
3   The gear selectors are in the upper crankcase half.
4   CB 400F models: Remove the neutral indicator switch retaining plate after unscrewing the cross-head screw, and pull out the switch. Flatten the tab washer on the bolt securing the retaining plate for the selector drum bearing. Unscrew the bolt, remove the retaining plate and pull out the selector drum, (there is a tapped hole in the end to assist).

20.1 Remove the connecting rod

21.4a Pull out neutral indicator switch (CB 400F only)

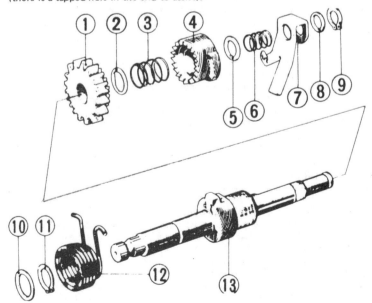

Fig. 1.11b. Kickstart assembly, CB 550 models

1   Kickstart pinion
2   Thrust washer
3   Starter pinion spring
4   Kickstart ratchet
5   Thrust washer
6   Kickstart ratchet spring
7   Ratchet guide plate
8   Chain guide thrust washer
9   Circlip
10 Washer
11 Circlip
12 Kickstart return spring
13 Kickstart spindle

CB 550 models: Remove the neutral indicator switch cam after unscrewing the crosshead screw in the centre. Unscrew the two hexagonal bolts securing the selector drum bearing retaining plate. Extract the spring pin and guide pin from the selector fork mounted on the selector drum. Withdraw the drum, holding the selector fork.

5   The drum bearing retaining plate also retains the selector fork shaft. Screw a 6 mm bolt into the end of the shaft, and pull it out holding the selector forks. Put the forks back on the shaft in the order they came off. On the CB 400F model, the forks are marked 'R', 'L' and 'C' for right, left and centre.

6   Flatten the tab washer on the starter gear shaft bolt, and unscrew the bolt. Screw a long 6 mm bolt into the end of the shaft, through the hole in the crankcase. Pull out the shaft and remove the starter gear.

22 Dismantling the engine/gearbox unit: removing the kickstart shaft, CB 550 models only

1   The kickstart shaft is in the bottom crankcase half, so that the gear assemblies need not be disturbed.

2   Remove the washer from the splined end of the shaft, and extract the circlip. Unhook and withdraw the return spring. Extract the circlip from the inner end of the shaft, and withdraw the shaft and mechanism.

23 Examination and renovation: general

1   Before examining the parts of the dismantled engine unit for wear, it is essential that they should be cleaned thoroughly. Use a paraffin/petrol mix to remove all traces of old oil and sludge that may have accumulated within the engine.

2   Examine the crankcase castings for cracks or other signs of damage. If a crack is discovered, it will require professional repair, or renewal.

3   Carefully examine each part to determine the extent of wear, checking with the tolerance figures listed in the Specifications Section of this Chapter. If there is any question of doubt, play safe and renew.

4   Use a clean lint free rag for cleaning and drying the components. This will obviate the risk of small particles obstructing the internal oilways, causing the lubrication failure.

24 Big-end bearings: examination and renovation

1   Big-end failure is invariably indicated by a pronounced knock from the crankcase. The knock will become progressively worse, accompanied by vibration. It is essential that the bearings are replaced as soon as possible, since the oil pressure will be reduced, and damage caused to other parts of the engine.

2   Big-end wear can be assessed before separating the crankcase, by trying to pull and push vertically on the connecting rods, when they are top dead centre. There should be no discernable play.

3   The big-ends have shell type bearings. Examine the bearing surfaces after removing the bearing caps, it is not necessary to remove the shell. If the bearing surfaces are badly scuffed or scored, the shells will have to be renewed. All the big-end bearing shells must be renewed at the same time.

4   Examine also the crankshaft journals for damage. Measure the journal diameter with a micrometer. Measure in several positions to check for ovality. If a journal is outside the service limit, the crankshaft must be renewed.

5   To select the bearing shell required, the crankshaft journal diameter must be measured. Check the connecting rod code number (stamped on the side of the big-end), and select the required shell from the table. The code is marked on the outside of the shell, as a paint dot or stamped letter. The letter stamped on the connecting rod is the weight code.

6   Fit the connecting rod to the crankshaft and measure the side play with feeler gauges. Renew if outside the service limit.

25 Crankshaft and main bearings: examination and renovation

1   Examine the main bearing shells and crankshaft journals for scuffing or scoring. If the shells are badly damaged, they must be renewed. All bearing shells must be renewed at the same time. It is important that the main bearings be overhauled immediately wear is discovered, since increased clearances will reduce oil pressure and possibly cause damage to other parts of the engine.

2   Measure the crankshaft bearing journals. Measure in several positions to check for ovality. If a journal is worn beyond the service limit, the crankshaft must be renewed.

3   With the crankshaft supported at each end in vee-blocks, measure the crankshaft runout at the centre main bearing, using a dial gauge. If the runout exceeds the service limit, the crankshaft must be straightened by a specialist, or renewed.

4   Check that there are no sharp edges on either end of the crankshaft, which may damage the oil seal as it is put on.

5   The main bearing shells must be selected as follows: Remove the bearing shells and clamp together the crankcases. Torque the bolts to the specified figure. Measure the inside diameter of each bearing houring vertically. Measure the diameter of each bearing journal and select the required bearing shell from the table. Note; the lower crankcase, and crankshaft are marked with the letters or numbers by the factory. These are production codes only. The shells are marked with a dot of paint, or a stamped letter on the back of the shell.

6   Check the security of the steel balls which plug the oil passageways at each end of the crankshaft.

26 Connecting rods: examination and renovation

1   It is unlikely that any of the connecting rods will bend during normal usage, unless an unusual occurrence such as a dropped valve has caused the engine to lock. Carelessness when removing a tight gudgeon pin can also give rise to a similar problem. It is not advisable to straighten a bent connecting rod; renewal is the only satisfactory solution.

2   The small end eye of the connecting rod is unbushed and it will be necessary to renew the connecting rod if the gudgeon pin becomes a slack fit. Always check that the oil hole in the small end eye is not blocked since if the oil supply is cut off, the bearing surfaces will wear very rapidly. Measure the inside diameter of the small end with a micrometer.

27 Oil seals: examination and renovation

1   Check all oil seals for scratches or damage to the sealing lip, for a damaged garter spring or for hardening.

2   It is difficult to give any firm recommendation, other than that the main seals (crankshaft and gearshaft) should be renewed whenever the engine is dismantled, as a precaution. If a seal is leaking, it should of course be renewed without question.

28 Cylinder bores: examination and renovation

1   The usual indication of badly worn cylinder bores and pistons is excessive smoking from the exhausts, and piston slap, a metallic rattle that occurs when there is little or no load on the engine. If the top of the bore of the cylinder block is examined carefully, it will be found that there is a ridge on the thrust side, the depth of which will vary according to the wear that has taken place. This marks the limit of travel of the top piston ring.

2   Measure the bore diameter just below the ridge, using an internal micrometer. Compare this reading with the diameter at

21.4b Remove selector drum bearing retaining plate (CB 400F shown)

21.5 Extract the selector fork shaft (CB 400F shown)

21.6a Unscrew starter gear shaft bolt ...

21.6b ... and withdraw the shaft

24.5 Bearing code is stamped on shell

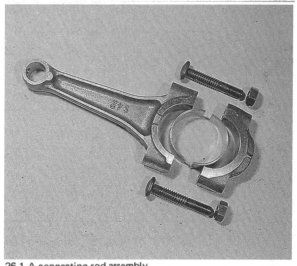

26.1 A connecting rod assembly

Crankcase allowance \ Crankshaft journal allowance		1	$32\phi \begin{smallmatrix}+0\\-0.010\end{smallmatrix}$	2	$32\phi \begin{smallmatrix}-0.010\\-0.020\end{smallmatrix}$	3	$32\phi \begin{smallmatrix}-0.020\\-0.030\end{smallmatrix}$
		Bearing mark (color)	Clearance (μ)	Bearing mark (color)	Clearance (μ)	Bearing mark (color)	Clearance (μ)
A	$35\phi \begin{smallmatrix}+0.008\\-0\end{smallmatrix}$	D (YELLOW)	20~46	C (GREEN)	22~48	B (BROWN)	24~50
B	$35\phi \begin{smallmatrix}+0.016\\+0.008\end{smallmatrix}$	C (GREEN)	20~46	B (BROWN)	22~48	A (BLACK)	24~50
C	$35\phi \begin{smallmatrix}+0.024\\+0.016\end{smallmatrix}$	B (BROWN)	20~46	A (BLACK)	22~48	AA (BLUE)	24~50

Fig. 1.12a. Selection of crankshaft bearing shells - CB 400F model

Connecting rod allowance \ Crank pin allowance		A	$32\phi \begin{smallmatrix}+0\\-0.010\end{smallmatrix}$	B	$32\phi \begin{smallmatrix}-0.010\\-0.020\end{smallmatrix}$	C	$32\phi \begin{smallmatrix}-0.020\\-0.030\end{smallmatrix}$
		Bearing mark (color)	Clearance (μ)	Bearing mark (color)	Clearance (μ)	Bearing mark (color)	Clearance (μ)
1	$35\phi \begin{smallmatrix}+0.008\\-0\end{smallmatrix}$	E (RED)	18~44	D (YELLOW)	20~46	C (GREEN)	22~48
2	$35\phi \begin{smallmatrix}+0.016\\+0.008\end{smallmatrix}$	D (YELLOW)	18~44	C (GREEN)	20~46	B (BROWN)	22~48
3	$35\phi \begin{smallmatrix}+0.024\\+0.016\end{smallmatrix}$	C (GREEN)	18~44	B (BROWN)	20~46	A (BLACK)	22~48

Fig. 1.12b. Selection of big-end bearing shells - CB 400F model

Crankcase allowance \ Crankshaft journal allowance		1	$33\phi \begin{smallmatrix}+0\\-0.010\end{smallmatrix}$	2	$33\phi \begin{smallmatrix}-0.010\\-0.020\end{smallmatrix}$
		Bearing mark (color)	Clearance (μ)	Bearing mark (color)	Clearance (μ)
C	$36\phi \begin{smallmatrix}+0.006\\-0.002\end{smallmatrix}$	B (BROWN)	20—46	A (BLACK)	22—48
B	$36\phi \begin{smallmatrix}-0.010\\-0.002\end{smallmatrix}$	C (GREEN)	20—46	B (BROWN)	22—48
A	$36\phi \begin{smallmatrix}-0.018\\-0.010\end{smallmatrix}$	D (YELLOW)	20—46	C (GREEN)	22—48

Fig. 1.13a. Selection of crankshaft bearing shells - CB 550 models

Connecting rod allowance \ Crank pin allowance		A	$35\phi \begin{smallmatrix}+0.006\\-0.004\end{smallmatrix}$	B	$35\phi \begin{smallmatrix}-0.014\\-0.004\end{smallmatrix}$
		Bearing mark (color)	Clearance (μ)	Bearing mark (color)	Clearance (μ)
3	$38\phi \begin{smallmatrix}+0.021\\+0.014\end{smallmatrix}$	B (BROWN)	18—43	A (BLACK)	20—45
2	$38\phi \begin{smallmatrix}+0.014\\+0.007\end{smallmatrix}$	C (GREEN)	19—44	B (BROWN)	21—46
1	$38\phi \begin{smallmatrix}+0.007\\-0\end{smallmatrix}$	D (YELLOW)	20—45	C (GREEN)	22—47

Fig. 1.13b. Selection of big-end bearing shells - CB 550 models

the bottom of the cylinder bore, which has not been subjected to wear. If the difference in readings exceeds 0.12 mm (0.005 in) the cylinder should be rebored and fitted with an oversize piston and rings.

3 If an internal micrometer is not available, the amount of cylinder bore wear can be measured by inserting the piston without rings so that it is approximately ¾ inch from the top of the bore. If it is possible to insert a 0.005 inch feeler gauge between the piston and the cylinder wall on the thrust side of the piston, remedial action must be taken.

4 Honda can supply pistons in four oversizes: 0.25 mm (0.010 inch), 0.050 mm (0.020 inch), 0.075 mm (0.030 inch) and 1.0 mm (0.040 inch). The 1 mm oversize is the safe limit to which the cylinder bores can be increased in diameter.

5 Check that the surface of the cylinder bores is free from score marks or other damage that may have resulted from an earlier engine seizure or a displaced gudgeon pin. A rebore will be necessary to remove any deep scores, irrespective of the amount of bore wear that has taken place, otherwise a compression leak will occur.

6 Make sure the external cooling fins of the cylinder block are not clogged with oil or road dirt, which will prevent the free flow of air and cause the engine to overheat.

29 Pistons and rings: examination and renovation

1 If a rebore is necessary, ignore this section, since new components will be fitted.

2 If a rebore is not considered necessary, examine each piston closely. Reject pistons that are scored or badly discoloured as the result of exhaust gases by-passing the rings.

3 Remove all carbon from the piston crowns, using a blunt scraper, which will not damage the surface of the piston. Clean away all carbon deposits from the valve cutaways and finish off with metal polish so that a clean, shining surface is achieved. Carbon will not adhere so readily to a polished surface.

4 Check that the gudgeon pin bosses are not worn or the circlip grooves damaged. Check that the piston ring grooves are not enlarged. Slide float should not exceed 0.007 inch (0.18 mm) compression rings or 0.005 inch (0.12 mm) oil control rings.

5 Piston ring wear can be measured by inserting the rings in the bore from the top and pushing them down with the base of the piston so that they are square in the bore and about 1½ inches down. If the end gap exceeds 0.027 inch (0.7 mm) on all rings, renewal is necessary.

6 Check that there is no build up of carbon on the inside surface of the rings or in the grooves of the pistons. Any build up should be removed by careful scraping. An old, broken ring is useful for this.

7 The piston crowns will show whether the engine has been rebored on some previous occasion. All oversize pistons have the rebore size stamped on the crown. This information is essential when ordering replacement piston rings.

8 The letter stamped on the face of a piston ring indicates the top of the ring, also the manufacturer. Rings from one manufacturer only should be used on any one piston.

30 Cylinder head and valves: dismantling, examination and renovation

1 Remove the carbon from the combustion chambers before removing the valves. Use a blunt scraper, which will not damage the surface and finish with metal polish.

2 A valve spring compressor is required to remove the valves. Compress the springs and remove the valve collets. Relax the springs and remove the valve spring collar valve spring, valve spring seats and valve. Keep these components together in a set. Each valve must be replaced in its original location.

3 Clean the carbon from the inlet and exhaust ports and from the head of the valve.

4 Before attending to the valve seats, check the valve guide and stem wear. If it exceeds the stated limit, the guide must be renewed. Drive out the guide from the combustion chamber side, using a drift of the correct diameter. The new guide must be reamered to the correct diameter after fitting. The valve seat must be re-cut after renewing the guide.

5 Check also that the valve stem is not bent, especially if the engine has been over-revved. If it is bent, the valve must be renewed.

6 The valves must be ground in to provide a gas-tight seal, during normal overhaul, or after recutting the seat or renewing the valve.

7 Valve grinding is a simple, but laborious task. Smear grinding paste on the valve seat and attach a suction grinding tool to the valve head. Oil the valve stem. Rotate the valve in both directions, lifting it occasionally and turning it through 90°. Start with coarse paste if the seats are badly pitted and continue with fine paste until there is an unbroken matt grey ring on each seat and valve. After many re-grinds, the valve seat may become pocketed, when it should be re-cut. Wipe off very carefully all traces of

30.2 Compress the valve springs, and remove the valve collets

30.11a Fit spring seat and oil seal

grinding paste. If any remains in the engine, it will cause very
rapid wear!

8   Valve sealing may be checked by installing the valve and
spring and pouring paraffin down the port. Check that none leaks
past the valve seat.

9   Check that the valve collets seat well in their grooves. Also
check the valve spring collar. Renew any defective part.

10 Check the free length of the valve springs. Renew when below
the service limit. Alternatively, compare with new springs.

11 Fit new oil seals to each valve guide on the CB 550 model, or
each inlet guide on the CB 400F model after checking that the
valve spring seats are in place. Oil the valve stem, and replace the
valve. Fit both valve springs with the close coils next to the
cylinder head, followed by the valve spring collar. Compress the
springs and fit the valve collets. Relax the springs, ensuring that
the collets remain in position. Seat the collets firmly with a few
blows squarely on the valve stem with a soft hammer. The
cylinder head should **not** be resting flat on the bench when
doing this.

12 Check the cylinder head joint face for flatness. If the face has
more than 0.3 mm (0.011 inch) total bow, it must be machined
flat, or the head renewed. Cylinder head joint warping is generally
due to incorrect tightening of the holding down bolts.

13 If the engine has been running weak, check the inlet manifold
adaptors for leaks. Make sure the vacuum gauge plugs are
sealing well. On the CB 500 models, the two pairs of adaptors
are clamped to the head with four nuts. On the CB 400F model,
each individual adaptor has two screws. After removing the
adaptors, check their 'O' ring seals. Check that the joint faces
are not warped; this will be caused by over-tightening.

31 Rocker spindles and arms: examination and renovation

1   Remove the rocker spindle from the rocker cover. On the
CB 400F model the rocker spindle cap must be unscrewed first.
Withdraw the spindle after screwing a suitable bolt into its end
(10 mm on CB 400F models, 6 mm on CB 550 models).

2   On the CB 400F model, note the position of the rocker arm
springs. Remove the rocker arms.

3   Check the rocker arm clearance when assembled on the
spindle. Examine the ends of the rocker arms for damage. Also
examine the tappet. Renew any defective part.

4   Oil all parts liberally when reassembling. The rocker arms
must move freely.

32 Camshaft, cam chain and tensioner: examination and renovation

1   Examine the camshaft bearings for scuffing or scoring.

2   Examine the camshaft and cams. Measure the height of each

cam. Renew the camshaft if any is worn below the service limit.
If the flanks of the cams are scored or damaged it is also
necessary to renew the camshaft.

3   It must be emphasised that most cases of rapid camshaft wear
can be traced to failure to change the engine oil at the
recommended time. It is clearly false economy to extend the
period between oil changes.

4   The camshaft drive chain can be checked for wear by
stretching it tight, and measuring the length of the chain. If it
measures more than 6 mm per 300 mm (¼ in per foot) of the
original length, it should be renewed.

5   The cam chain guide and tensioner should not be worn below
the service limit.

33 Cam chain sprockets and rev-counter drive: examination and renovation

1   The upper camshaft chain sprocket is bolted to the camshaft
and in consequence is easily renewable if the teeth become
hooked, worn, chipped or broken. The lower sprocket is integral
with the crankshaft and if any of these defects are evident, the
complete crankshaft assembly must be renewed. Fortunately,
this drastic course of action is rarely necessary since the parts
concerned are fully enclosed and well lubricated, working under
ideal conditions.

2   If the sprockets are renewed, the chain should be renewed at
the same time. It is bad practice to run old and new parts
together since the rate of wear will be accelerated.

3   The worm drive to the tachometer is an integral part of the
camshaft which meshes with a pinion attached to the cylinder
head cover. If the worm is damaged or badly worn, it will be
necessary to renew the camshaft complete.

34 Primary chain and guides: examination and renovation

1   Examine the primary chain for wear or damage. Also check
the sprockets.

2   Primary chain guides are fitted to the CB 400F model only.
The guides are bolted to the upper and lower crankcases. Check
that they are not worn below the service limit. When fitting
guides, the incised dot must be at the rear end of the guide.

35 Clutch assembly: examination and renovation

1   The seven clutch friction plates have a bonded lining. When
the serviceable limit of wear is reached, clutch slip will occur.
Measure with a vernier caliper and renew the plates when

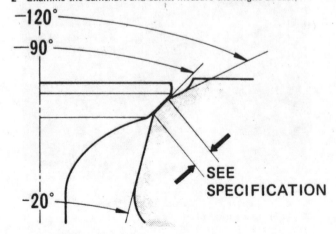

Fig. 1.14. Valve seat dimensions

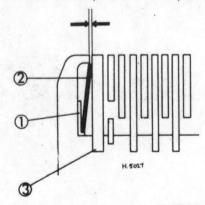

Fig. 1.15. Clutch plate 'B' clearance, CB 400F model only

1   Spring seat   3   Clutch plate 'B'
2   Spring

30.11b The close coils go next to the cylinder head

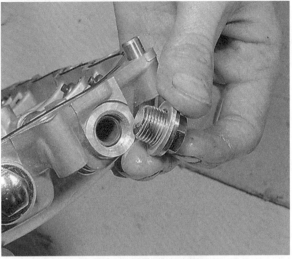

31.1a Remove the rocker spindle cap (CB 400F) ...

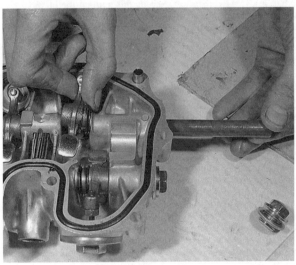

31.1b ... and withdraw the rocker spindles

34.2 The dot must be at the rear of the guide (CB 400F only)

35.5 CB 400F clutch centre has an additional plain plate

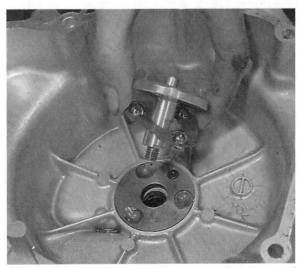

36.1 Remove the clutch thrust pad and ball cage (CB 400F only)

37.1a Remove needle bearing from starter gear

37.1b Pull off starter gear

37.2 Starter clutch roller and plunger displaced

necessary - they cannot be relined. Always renew the plates as a complete set, never singly.

2   Another factor which can promote clutch slip is clutch springs which have taken a permanent set and lost some of their original tension. If the serviceable limit has been reached, replacement of all four springs as a set is essential.

3   Check that none of the clutch plates, either plain or bonded is buckled. Renewal will be necessary if the plates do not lie flat, since it is difficult to straighten them with any measure of success.

4   Examine the clutch assembly for burrs on the edges of the protruding tongues of the inserted plates and/or slots worn in the edges of the outer drum with which they engage. Similar wear can occur between the inner tongues of the plain clutch plates and the slots in the clutch inner drum. Wear of this nature will cause clutch drag and other troubles, since the plates will become trapped and will not free fully when the clutch is withdrawn. A small amount of wear can be treated by dressing with a file; more extensive wear will necessitate renewal of the worn parts.

5   CB 400F models only: The clutch centre on the CB 400F models has an additional plain plate 'B', retained by a large circlip. Measure the clearance between the plate, and the clutch centre (see illustration). This should be 0.1 - 0.5 mm (0.004 - 0.020 in). If the dimension exceeds this, replace the plate. Assemble the spring and spring seat as shown.

6   CB 400F models only: Examine the clutch thrust bearing for damage to the balls or tracks, or for excessive radial play. Push it out of the spider if renewal is necessary.

7   Check that the clutch gear wheel is securely rivetted to the clutch drum, it may be possible to tighten the rivets, but renewal is the only satisfactory long term solution.

36 Clutch operating mechanism: examination and renovation

1   On the CB 400F model, a release mechanism quite different from that of the CB 550 models is employed. Remove the triangular clutch adjuster cover and the kickstart. Disconnect the clutch cable and remove the clutch cover. Remove the adjuster locknut, while preventing the adjuster screw from turning. Remove the washer, lever and spring. Pull out the thrust pad from inside the cover. Remove the three ball cage and cam plate, which locates on a dowel. Check all parts for wear or damage.

2   Little attention is likely to be needed to the mechanism of the CB 550 models, other than replacing a faulty oil seal or spring. Also check the clutch pushrod.

3   When reassembling, the lever with the slot in the shackle facing outwards, should point at the arrow on the cover. Position the slotted hole in the lever over the thrust pad shaft and replace the washer, nut and spring. If the oil seal should require renewing, it may be prised out from the outside of the cover, as may the kickstart shaft seal.

37 Primary drive and starter clutch: examination and renovation

1   Withdraw the needle roller bearing from the starter gear. On the CB 550 models there is an additional spacer, which may remain on the primary shaft, or in the starter gear. Pull off the starter gear from the starter clutch, taking care not to drop the clutch rollers.

2   The three rollers are pushed against a stop by spring loaded plungers. Ensure that the rollers are fitted correctly and move freely.

3   Pull the clutch out of the shock absorber in the sprocket. On the CB 550 models a circlip has to be removed first. The eight similar shock absorber rubbers have moulded arrows, pointing outwards. Refit the clutch with the vanes between each pair of rubbers. The shock absorber should be quite tight, with no rotary free-play. Do not dismantle further.

4   Check the condition of the starter gear teeth. Push the starter gear spigot into the starter clutch, taking care not to displace

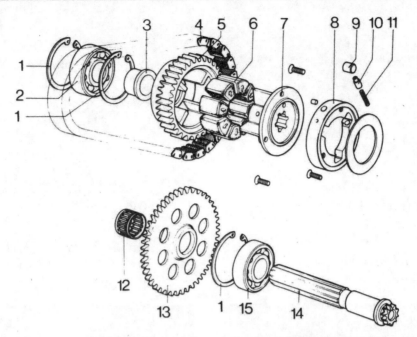

Fig. 1.16. Primary shaft and starter clutch (CB 400F model shown)

1	Circlips - 3 off	5	Primary chain sprocket	9	Roller - 3 off
2	Ball journal bearing	6	Rubber damper - 8 off	10	Plunger - 3 off
3	Flanged sleeve	7	Driven sprocket hub	11	Spring - 3 off
4	Primary chain	8	Clutch outer	12	Needle bearing

13	Starter driven gear
14	Primary shaft
15	Ball journal bearing

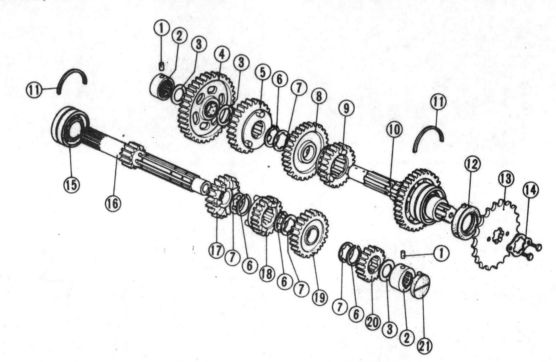

Fig. 1.17. Gearbox assembly, CB 400F model

1	Dowel - 2 off	6	Circlip - 4 off	12	Oil seal
2	Needle roller bearing - 2 off	7	Splined thrust washer - 4 off	13	Final drive sprocket (17T)
3	Thrust washer - 3 off	8	Countershaft third gear (30T)	14	Sprocket locking plate
4	Countershaft bottom gear (41T)	9	Countershaft top gear (26T)	15	Ball journal bearing
5	Countershaft fifth gear (28T)	10	Countershaft	16	Mainshaft (24T)
		11	Bearing retaining ring - 2 off	17	Mainshaft fifth gear (29T)

18	Mainshaft third and fourth gear (24T and 27T)
19	Mainshaft top gear (30T)
20	Mainshaft second gear (20T)
21	Oil seal

37.3 Pull clutch out of shock absorber

38.1 Gear assembly, CB 400F. Mainshaft at top

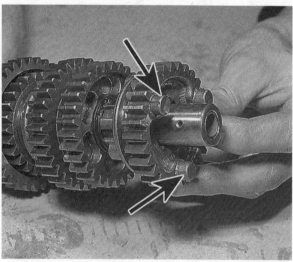

38.2a Check ends of the dogs are not rounded

38.2b Freely rotating gears should have barely discernable radial play

38.5 Take care when removing or replacing circlips

38.6a Do not omit the many thrust washers

the rollers. Refit the needle roller bearing, followed by the spacer on the CB 550 models.

5   Check the primary shaft ball journals. The drive side bearing on the CB 550 models remains on the shaft, remove the drive pinion circlip, pinion, spacer and pull off the bearing if it is to be renewed. The oil pump side bearing remains in the crankcase, retained by circlips on the CB 400F model. Check the bearings in-situ. They should rotate freely, without roughness or excessive play.

38 Gearshafts: examination and renovation

1   It is possible to check the condition of gear components without dismantling the shaft assemblies.

2   Check the gear teeth for wear or damage, and that the sliding gears slide freely. Check that the ends of the dogs on the sliding gears are not rounded, or damaged. The freely rotating gears should have barely discernable radial play. One or two of the gears (depending on model) have pressed-in bushes, although these are not shown as separate items in the parts list.

3   If any part has to be renewed, dismantle the shafts using the illustrations as a guide. Note carefully the order of components. The CB 400F model, has an extra gear on each shaft.

4   Check that the ball journal bearings revolve freely without roughness or excessive radial play. The component parts of the needle bearings can be examined separately. Check that the rollers and roller tracks are undamaged.

5   Make sure that the shafts have no sharp edges to damage the oil seals as they are put on.

6   When reassembling gear shafts, take care not to distort the circlips. The best way to install them, is by pushing them along the shaft with a tube having a faced-off square end.

7   If the gears have been removed from the shafts, note the following points on assembly:

CB 550 models: There is an 'O' ring seal over the countershaft, behind the ball journal bearing. The thrust washer having six splines goes between the second and third countershaft gears, and is retained by the splined lock washer. Push the lock washer along the shaft and turn the thrust washer to engage the tongues on the lock washer. Take care that the lock washer is not displaced while the other components are being assembled. Do not omit the plain thrust washer, between mainshaft top gear and the needle bearing.

38.6b Push lockwasher along the shaft ...

38.6c ... and turn thrust washer to engage tongues

38.6d Don't omit thrust washer between last gear and needle bearing

38.6e Align oil holes in shaft and bush (CB 400F only)

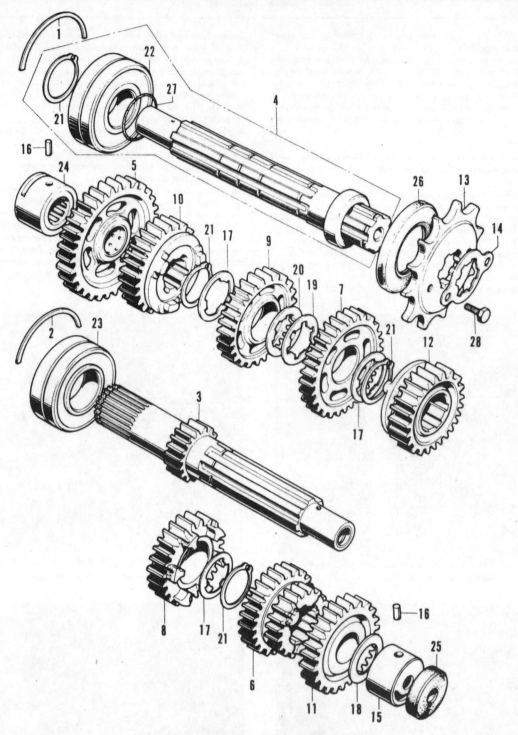

Fig. 1.18. Gearbox assembly, CB 550 models

1 Bearing retaining ring
2 Bearing retaining ring
3 Mainshaft (24T)
4 Countershaft
5 Countershaft bottom gear (40T)
6 Mainshaft second and third gears (22T, 26T)
7 Countershaft second gear (36T)

8 Mainshaft fourth gear
9 Countershaft fourth gear (33T)
10 Countershaft fourth gear (29T)
11 Mainshaft top gear (30T)
12 Countershaft top gear (27T)

13 Final drive sprocket (17T)
14 Sprocket fixing plate
15 Needle roller bearing
16 Dowel
17 Thrust washer - 3 off
18 Thrust washer
19 Thrust washer
20 Lockwasher

21 Circlip - 4 off
22 Ball journal bearing
23 Ball journal bearing
24 Needle roller bearing
25 Oil seal
26 Oil seal
27 O-ring seal
28 Sprocket fixing bolt - 2 off

CB 400F model: There is a splined thrust washer between the third and fourth countershaft gears, retained by a lockwasher as described above. The oil hole in the splined bush on the mainshaft should be aligned with the oil hole in the shaft.

8 If the ball journal bearings are being renewed, they must come from the same source as the originals, since the locating ring grooves may differ between manufacturers.

9 Replace both shafts in the upper crankcase and check that all gears engage correctly. Measure the backlash between gears. This requires the use of a dial gauge; Check that there is clearance between the ends of the dogs on adjacent gears.

39 Kickstart mechanism: examination and renovation

1 The kickstart shaft and mechanism on the CB 400F model can be withdrawn without separating the crankcases, unlike the CB 550 models. In either case, the return spring can be renewed without dismantling the engine.

2 Remove the kickstart and clutch cover. CB 550 models only: remove the washer and circlip from the end of the kickstart shaft.

3 Unhook the return spring from its crankcase location, and pull it off of the shaft. Engage the new spring with the location on the shaft and wind the spring clockwise until it will engage with the crankcase location.

4 When the shaft is removed, dismantle the ratchet mechanism. Check all parts for wear or damage, any of which is unlikely, since the electric starter will have been used generally. The most important item, which must be undamaged, is the ratchet. If this can slip, there will be rather painful results.

40 Gear selectors: examination and renovation

1 Examine all parts of the gear selector mechanism carefully for wear. Pay particular attention to the selector fork ends, and the pins and their slots in the selector drum.

2 If there are any signs of scuffing, or overheating (indicated by blueing of the fork ends), there is mis-alignment which must be rectified. Check the width of the fork ends.

3 The selector fork pins must slide freely in their grooves, without side play. Examine the grooves, especially where they change direction.

4 Check the selector drum bearing for damaged balls or tracks or excessive radial play.

5 Check all parts of the gearchange linkage. Pay particular attention to the ends of the gear shift lever, and the positive stopper on the CB 400F model. Also check the springs in the mechanism, especially the gearlever hairpin return springs. If these are weak, it will make gear selection difficult. If the machine has covered a high mileage, it is worthwhile to replace the springs, as a precaution against failure.

41 Engine and gearbox reassembly: general

1 Before reassembly is commenced, engine and gearbox components should be thoroughly clean and placed close to the working area.

2 Make sure all traces of old gaskets have been removed and that the mating surfaces are clean and undamaged. One of the best ways to remove old gasket cement, which is needed only on the crankcase and cover joints, is to use a rag soaked in methylated spirit. This acts as a solvent and will ensure the cement is removed without resort to scraping and the consequent risk of damage.

3 Gather all the necessary tools and have available an oil can filled with clean engine oil. Make sure that all new gaskets and oil seals are available; there is nothing more frustrating than having to stop in the middle of a reassembly sequence because a vital gasket or replacement has been overlooked.

4 Make sure the reassembly area is clean and well lit, with adequate working space. Refer to the torque and clearance

settings wherever they are given. Many of the smaller bolts are easily sheared if they are over-tightened. Always use the correct size screwdriver bit for the crosshead screws and NEVER an ordinary screwdriver or punch.

42 Reassembling the engine/gearbox unit: rebuilding the gearbox

1 Reassemble the selector forks and drum in reverse order to dismantling. Make sure the selector forks are put back in the correct position.

2 On the CB 550 models, insert the fork guide pin and engage it with the central slot in the drum. Insert the guide pin clip.

3 Place the selector drum in neutral, (CB 400F model: neutral indicator contact in view through the switch hole).

4 Replace the starter gear. Engage the end of the retaining bolt with the groove in the end of the shaft. The smaller gear pinion goes on the oil pump side.

5 Replace the neutral indicator switch. CB 550 models only: engage the switch cam with the notch in the end of the selector drum.

6 Fit the mainshaft bearing locating ring in the upper crankcase

42.1 Gear selector assembly (CB 400F only)

42.6a Fit bearing locating ring

half. Turn the mainshaft needle bearing so that its dowel hole is downwards. Place the mainshaft firmly in its housing, locating the bearing ring in its groove, and the needle bearing on its dowel. When correctly fitted, the inscribed line on the needle bearing should align with the crankcase joint face. The mainshaft sliding dog should engage with the centre selector fork.

7   Fit a new oil seal, with its dowel in the hole in the crankcase. The embossed line on the seal should align with the crankcase joint.

8   Put a new seal on the left-hand end of the countershaft, closed side facing outwards and integral dowel downwards. Make sure that the bearing locating ring is in place in the top crankcase half. Turn the needle bearing so the dowel hole is downwards. Place the shaft firmly into its housing, meshing the gears, and engaging the bearing locating ring and the oil seal dowel. Engage the needle bearing dowel; when correctly fitted, the line engraved on the bearing should align with the crankcase joint. The sliding dogs should engage with the two outer selector forks. The arrow embossed on the oil seal should face forwards, horizontally. The 'F' indicates 'front'.

9   Spin the shafts to ensure that everything turns freely. Turn the selector drum to engage each gear in turn, then return to neutral. Liberally oil the gears and bearings.

10 Make sure the oil guides are secure in the crankcase (CB 550 models only), they should not have needed to be removed. If one becomes loose, considerable damage will result.

43 Reassembling the engine/gearbox unit: installing the crankshaft

1   Fit the big-end bearing shells, if these have been removed. Ensure that the keys locate securely in their notches. Liberally oil the bearings.

2   Assemble each connecting rod onto its respective bearing journal. Both big-end shell keys **must** be positioned at the front (exhaust) of the crankshaft.

3   The washer faces of the big-end nuts must be adjacent to the big-end cap. Torque the nuts to the correct figure.

4   Fit the main bearing shells, if these have been removed. Ensure that the keys locate securely in their notches. Liberally oil the bearings.

5   Pass the primary and cam chains over the crankshaft and hook them onto their respective sprockets.

6   Place new oil seals on each end of the crankshaft. The alternator end is the larger. The closed side of the seals must face outwards. Do not refit the old seals, since a further stripdown will be needed if they fail.

7   Place the crankshaft firmly into the crankcase, lowering the cam chain and connecting rods down their respective holes. Make sure that the crankshaft oil seals are located correctly in their grooves.

8   CB 400F model only: Install the cam chain tensioner adjuster with the raised mark (which is in line with a flat on the shaft) facing forwards. Compress the adjuster against spring pressure, and tighten the adjuster bolt. Replace the semi-circular tensioner arm.

44 Reassembling the engine/gearbox unit: crankcase

1   CB 550 models only: Refit the kickstart shaft into the lower crankcase half. Insert the shaft through the ratchet assembly, and replace the inner circlip. Fit the return spring and replace the outer circlip. Do not omit the washer after the circlip.

2   Hook the input shaft sprocket through the primary chain and mesh the starter gears. On the CB 400F model the starter pinion is on the oil pump side. On the CB 550 models it is on the clutch side.

3   Clean the crankcase joint faces and replace the two dowels. Coat the joint faces with gasket cement, and bring them together. Tap gently to seat the bearings and make sure that the joint closes all round.

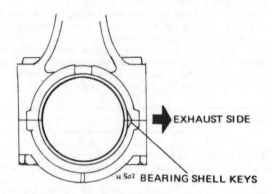

Fig. 1.19. Fit the keys on the big-end shells to the front

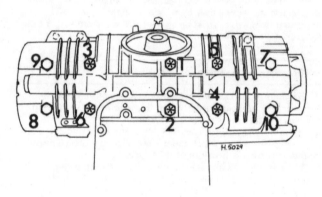

Fig. 1.20a. Crankcase bolt tightening sequence, CB 400F models (* bolts marked 'P')

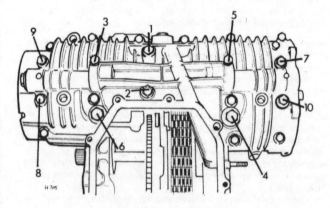

Fig. 1.20b. Crankcase bolt tightening sequence, CB 550 models (* bolts marked '9')

42.6b Place mainshaft in its housing

42.6c Inscribed line on bearing should align with joint face

42.7 Fit new mainshaft oil seal

42.8a Put new seal on countershaft

42.8b Place countershaft in its housing

42.8c Mesh all gears (CB 400F, shown in neutral)

42.10 Make sure the oil guides are secure (CB 550's only)

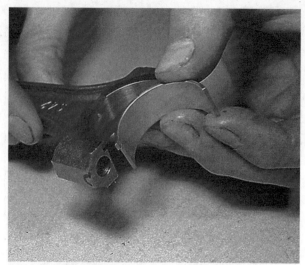

43.1 Fit new big end shells ...

43.4 ... and main bearing shells. Note locating key

43.5 Hook the chains over crankshaft ...

43.6 ... and put new oil seals on each end

43.8 Fit cam chain tensioner adjuster. Note raised mark (CB 400F only)

44.1 Replace the kickstart mechanism and shaft (CB 550 models only)

44.2a Fit sprocket onto primary chain on CB 550 models ...

44.2b ... and on CB 400F model note starter gear is on different side

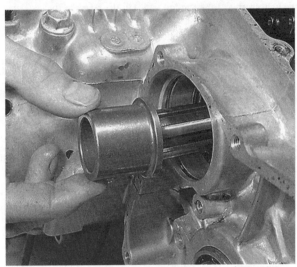

44.6a Flange on spacer goes next to starter clutch (CB 400F model)

44.6b Fit the ball journal bearing ...

44.6c ... and circlip...

4 Replace the ten 8 mm bearing cap bolts, and tighten them down uniformly in the sequences shown in illustrations 1.20a or b. CB 400F models: the inner six bolts are longer and have 'P' stamped on the head. CB 550 models: the two bolts stamped '9' on the head should be fitted at the alternator end.

5 Insert and gradually torque tighten down the remaining bolts, not forgetting the one inside the sump on the CB 550 models.

6 CB 400F models: Insert the input shaft from the oil pump side passing it through the flanged spacer. The flange must abut the starter clutch. Fit the ball journal bearing and circlip, followed by plain spacer and gear pinion. The pinion goes on either way round. The dished washer is put on with 'outside', outside. Torque tighten the bolt after clutch assembly, when rotation can be prevented.
CB 550 models: Make sure that the circlip and thrust washer are fitted to the input shaft. Insert the shaft into the starter clutch splines, from the clutch side. Take care not to displace the needle rollers and spacer from the starter gear. Tap the shaft fully home.

7 Turn the engine the right way up, and fit the bolts in the top. CB 550 models: The two 8 mm bolts stamped '8' on the head should be fitted on the clutch side of the casing.

8 CB 400F model only: Insert the kickstart shaft, engaging the hairpin end of the friction spring in the recess in the crankcase. Wind the return spring clockwise, and hook it into the crankcase.

45 Reassembling the engine/gearbox unit: gear selector mechanism

1 With the selector drum in neutral, replace the selector mechanism in reverse order to dismantling.

2 CB 400F model: Fit the neutral indexing (9) and selector drum indexing (4) levers, and hook their springs into the crankcase. Fit the 'Y' shaped lever (8), with the legs of its hairpin spring on either side of the spring post/pivot of the neutral indexing lever. Replace the selector cam with its dowel. When in neutral the indexing lever locates in the notch in the inner cam. The ends of the hairpin spring on the gear lever shaft arm go on either side of the tongue on the arm and the post in the crankcase.
CB 550 models: Fit the gear pedal shaft (16), with the legs of the hairpin spring on the lever, on either side of the post in the crankcase. Attach one of the tension springs between the selector drum indexing lever (11), and the arm of the neutral indexing lever (15). Hook the other spring between the neutral indexing lever and arm. Tighten the lever pivot bolt. Fit the bolt at the bottom end of the arm of the neutral indexing lever. Tighten both bolts in the retaining plate for the input shaft bearing. When in neutral, the roller on the neutral indexing lever should locate in the notch in the inner selector cam.

3 Check that the mechanism works correctly.

46 Reassembling the engine/gearbox unit: replacing the clutch

1 Build up the clutch as an assembly on the bench. The flanged sleeve must be in place in the clutch drum, flange outwards. Install the pressure plate, followed by the friction and plain plates alternately. The smaller friction plate goes on last, followed by the clutch centre - line up the splines on the plain plates.

2 Replace the thrust washer on the clutch shaft.

3 Holding the clutch assembly together, install it on the clutch shaft, engaging the splines and the pinion. CB 400F model: replace the dished washer, with 'outside' outside, tab washer and castellated nut. Note the tab washer locates on ears on the clutch centre. CB 550 models: Replace the shim and the circlip.

4 Replace the springs and spider, and tighten the bolts fully. CB 550 models only: Replace the clutch pushrod.

5 CB 400F model only: With the clutch replaced, the drive pinion bolt can be tightened by spragging the clutch.

6 CB 550 models only: When the complete clutch has been

44.6d ... followed by plain spacer ...

44.6e ... and pinion

44.6f 'OUTSIDE' on dished washer is self-explanatory

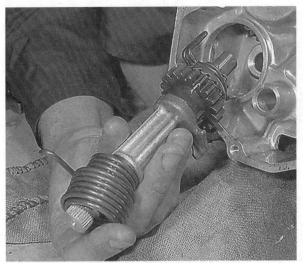

44.8 Fit kickstart shaft (CB 400F model only)

45.2a Hook spring into crankcase

45.2b Legs of hairpin spring rest either side of post

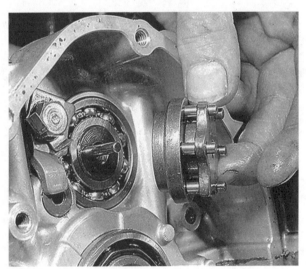

45.2c Note selector cam dowel

45.2d Legs of hairpin spring rest either side of post

46.1 The flange on the sleeve is on the outside

46.2 Replace the thrust washer

46.3a 'OUTSIDE' on dished washer faces is self-explanatory (CB 400 model only)

46.3b Replace the castellated nut (CB 400 model only)

47.2 Replace the alternator rotor

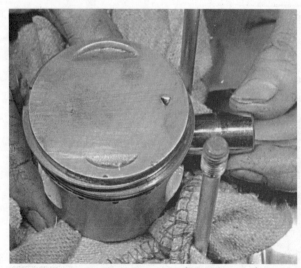

49.2a Arrow on the piston faces front (CB 550 models)

49.2b Valve cutaway marked 'IN' (CB 400F model only)

assembled, and the shim and circlip replaced, measure the clutch end float. This is best accomplished by using a dial gauge against the face of the clutch centre. If the float exceeds 0.1 mm (0.004 in) the shim behind the circlip should be replaced by a thicker one. Three different thicknesses are available.

47 Reassembling the engine/gearbox unit: replacing the oil pump, alternator and starter motor

1 Make sure all seals and dowels are in place and refit the oil pump. On the CB 400F model, fit the cable clip to the top bolt.
2 Clean the internal and external tapers. Replace the alternator rotor on the end of the crankshaft and torque tighten the fixing bolt.
3 Check the 'O' ring seal on the starter motor spigot. Push the motor into the crankcase, engaging the gears. Replace the two fixing bolts.
4 Replace the wiring harness (if removed), pushing the grommet into the cut-out in the crankcase. Connect up the neutral indicator switch, oil pressure switch and alternator.
5 Replace the alternator cover, using a new gasket.

48 Reassembling the engine/gearbox unit: replacing the contact breaker

1 Replace the automatic advance unit, inserting the dowel into the end of the crankshaft.
2 Replace the contact breaker assembly. Line up the centre punch marks made previously, and lightly tighten the screws. Fit the cable grommet into its cut-out.
3 Replace the cam fixing bolt. Note that the slots in the hexagonal washer locate on pegs on the cam.
4 Check the points gap and static ignition timing - see Chapter 4. (After completing engine reassembly, if necessary).
5 Replace the contact breaker cover and gasket.

49 Reassembling the engine/gearbox unit: replacing the pistons and cylinder block

1 Before replacing the pistons, place a clean rag in each crankcase mouth to prevent any displaced component from falling into the crankcase. It is only too easy to drop a circlip whilst it is being inserted into the piston boss, which would necessitate a further stripdown for its retrieval.
2 Fit the pistons in their original places, facing the right way. CB 550 models: The arrow stamped on the piston crown must face forwards.
CB 400F model: The valve cut-away marked 'in' must be on the inlet (rear) side. If the gudgeon pins are a tight fit, warm each piston first, to expand the metal. Do not forget to oil the gudgeon pins and the piston bosses before fitting.
3 Use new circlips, NEVER the originals. Check that the circlips have located correctly with the groove within each piston boss. A displaced circlip will cause severe engine damage. The circlips must overlap the screwdriver slot completely.
4 CB 400F: Install the two cylinder base dowels.
CB 550 models: Check that the oil restrictors are not blocked, and refit them with new 'O' ring seals. Fit new 'O' ring seals to the cylinder barrel spigots and a new cylinder base gasket. No gasket cement is necessary.
5 Position the piston rings so that their ends are 120° out of line with one another. A gap should not be positioned above a gudgeon pin boss, nor at front or rear. Fit piston ring clamps to the two centre pistons, raising them to top dead centre. Smear the cylinder bores with plenty of oil. Feed the camshaft drive chain upwards through the tunnel in the centre of the block and retain it in this position, then slide the block downward along the holding down studs until the piston ring clamps are displaced,

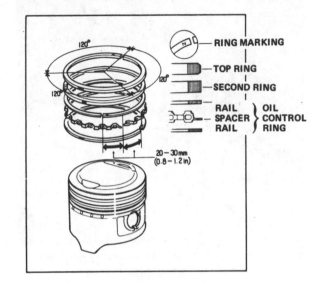

Fig. 1.21. Space the piston rings as shown

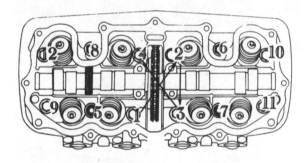

Fig. 1.22a. Cylinder head nut tightening sequence, CB 400F model (Nuts marked * are pillar nuts)

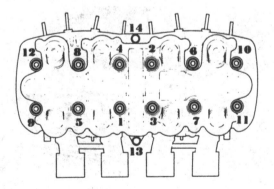

Fig. 1.22b. Cylinder head nut and bolt tightening sequence, CB 550F models

49.4 Replace oil restrictors and O-ring seals (CB 550 models)

49.5 Feed the two inner pistons into the block first

50.1a Replace the dowels with gaskets ...

50.1b ... and oil restrictors with gaskets (CB 400F model only)

50.1c Put a support under the cam chain

50.5 The inscribed line on sprocket aligns with joint face (CB 400F model only)

and the two centre pistons enter the cylinder bores correctly.
Remove the piston ring clamps and fit them to the two outer
pistons. Continue to move the cylinder block downwards
TOGETHER WITH THE TWO CENTRE PISTONS, so that the
outer pistons will rise from bottom dead centre as the crankshaft
rotates. Engage these pistons with their cylinder bores, then
remove the piston ring clamps and the rag padding from the
crankcase mouths before pushing the cylinder block firmly
onto the base gasket.

6   It is best to have assistance during this operation, otherwise
there is risk of piston ring breakage if the rings do not feed into
the bore in a satisfactory manner. It is possible to feed the rings
into the bores by hand but as this is a somewhat tedious task, it is
preferable to use piston ring clamps to simplify the operation.

7   Hook a support under the cam chain.
CB 400F model: Replace the cam chain tensioner at the rear of
the cylinder block and the cam chain guide at the front.
CB 500 models: Replace the cam chain tensioner in the back of
the cylinder barrel. Press it downwards by hand so that its
mounting stud protrudes through the boss in the rear of the
cylinder and install the 'O' ring and washer on the stud. Replace
and tighten the locknut. Replace the cam chain guide in the front
of the cylinder with the 'UP' mark facing rearwards.

8   If it is found difficult to replace the cam chain tensioner on
CB 550 models, raise the cylinder block about 20 mm (7/8 inch).
Take care that the piston rings do not come out of the bores
again.

**50 Reassembling the engine/gearbox unit: replacing the
cylinder head and rocker gear**

1   CB 550 models: Replace the two dowels at the rear of the
block, and the two 'O' ring seals at each end. CB 400F model:
Replace the two dowels at the rear of the block, also the two
dowels with tubular gaskets at the front, and the two oil
restrictors with tubular gaskets at each end (check that they are
not blocked). Put a new head gasket in position. On the CB
400F model, there is a small hole in the gasket at the left-hand
end. Check that the cam chain is still attached to the sprocket
on the crankshaft; feed the cam chain through the tunnel in the
centre of the cylinder head. Lower the cylinder head into

position down the holding down studs, whilst maintaining tension
on the cam chain.

2   CB 400F model: Oil the threads and replace the twelve nuts
in the same places as before, with the correct type of washer
under each. The two pillar nuts have the rubber bonded washer
under them. Tighten the nuts down uniformly in the sequence
shown (Fig. 1.22a). The final torque setting should be as given in
the Specifications.

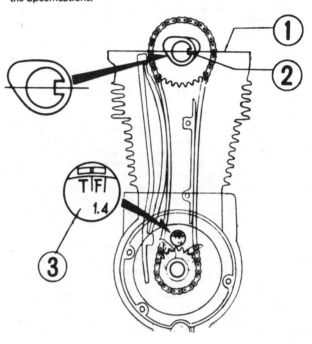

Fig. 1.23b. Alignment of valve timing marks, CB 550 models

1   Cylinder head joint face   3   Mark in timing window
2   Notch in end of camshaft

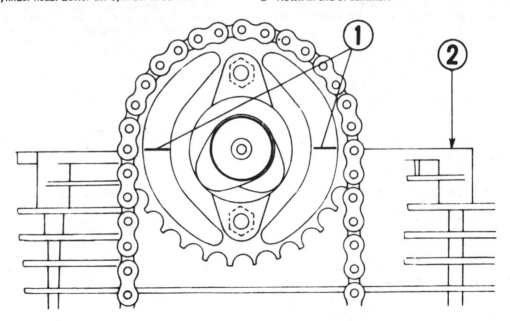

Fig. 1.23a. Alignment of valve timing marks, CB 400F model

1   Inscribed mark on camshaft sprocket   2   Cylinder head joint face

CB 550 models: Replace the twelve 8 mm nuts on the cylinder holding down studs and tighten them down in the sequence shown (Fig. 1.22b). Replace and tighten the two 6 mm flange bolts which are located forward and to the rear of the camshaft drive chain tunnel.

3 CB 550 models only: Fit the cam chain tensioner bolt and aluminium washer at the top.

4 Hold the cam chain sprocket and the cam chain together, and slide the camshaft through them from the right-hand side and set it on the bearings in the cylinder head.

5 Remove the contact breaker points cover and rotate the engine until pistons 1 and 4 are at top dead centre, then align EXACTLY the T mark for cylinders 1 and 4 with the static timing mark, as viewed through the aperture in the contact breaker baseplate. During this operation, maintain tension on the cam chain and check that it is still correctly engaged with the crankshaft sprocket.

6 CB 400F models: Turn the cam sprocket only until the inscribed line is parallel with the cylinder head flange face. Turn the camshaft to align the sprocket fixing holes, see Fig. 1.23a. CB 550 models: Rotate the camshaft so that the centre of the notch in the right-hand end of the shaft is aligned with the cylinder head flange face, see Fig. 1.23b.

7 Hook the cam chain onto the sprocket, and fit the sprocket onto its spigot, without disturbing any of the alignments. Replace one sprocket fixing bolt, turn the crankshaft 180° to replace the other. Failure to position the camshaft in this manner will result in incorrect valve timing.

8 CB 400F model: Replace the cam chain tensioner holder at the rear of the head. Hook the cam chain guide into its recess at the front. Replace the cam oil jet tubes, fitting the supporting grommets into their bosses (make sure the jet tubes are not blocked). CB 550 models: Replace the six sealing rubbers on the inner cylinder head nut.

9 Replace the two rocker cover dowels. Lubricate the camshaft and rocker gear. Fit a new rocker cover 'O' seal if necessary. Make sure all the tappets are slackened off before replacing the cover. CB 400F models: Torque tighten the bolts down gradually in the sequence shown in Fig. 1.24.

10 CB 550 models only: Replace the cylinder head side cover brackets. These are secured by a screw and two washers. The aluminium washer fits below the plate and the chrome washer fits on top. Install new 'O' rings on the dowel pins of the left and right-hand side covers and replace the side covers on the cylinder head.

11 Check and if necessary reset the valve clearances. Rotate the crankshaft clockwise using a spanner on the large hexagonal washer, until the 'T 1.4' mark aligns with the static timing mark, and No 1 cylinder is at TDC on compression (both valves closed). Now check and adjust if necessary, clearances on No 1 cylinder inlet and exhaust, No 2 cylinder exhaust, No 3 cylinder inlet. Turn the crankshaft forwards 360°, so that the 'T 1.4' mark aligns with the static mark when No 4 cylinder is at TDC on compression. This time check the clearances on No 4

cylinder inlet and exhaust, No 3 cylinder exhaust and No 2 cylinder inlet. In this way, the valves will be fully closed when checking clearances.

12 Adjust the tappets by slackening the locknut, and turning the adjuster. Tighten the locknut, and turning the adjuster. Tighten the locknut and recheck the clearance.

13 Replace the tappet covers. Do not forget to check the ignition timing as described in Chapter 3, before replacing the contact breaker cover.

14 Before the engine is replaced in the frame the cam chain adjustment should be checked as follows: CB 550 and 550F models: Remove the tappet covers from No1 cylinder inlet and exhaust. Remove the contact breaker cover. Rotate the crankshaft to align 'T 1.4' mark with the static timing mark, with No 1 cylinder on compression (both valves closed). Rotate the crankshaft a further 15° clockwise, ie until the spring post of the advance mechanism illustration is just to the right of the static timing mark.

15 The chain tension adjuster is at the base of the cylinder block, at the rear. Loosen the locknut, chain tension is then automatically adjusted. Retighten the locknut, and replace all parts.

16 The cam chain of the CB 400F models cannot be checked at this stage as adjustment must be made with the engine running. See Section 52.5.

---

51 Replacing the engine/gearbox unit in the frame

1 Rag wrapped round the front frame tubes will help to prevent damage as the engine is lifted in.

2 Follow the removal procedure in reverse. The two lower front engine mounting bolts **must** be the correct length and have a spring washer under their heads. If this is not so, the crankcase may be damaged. Clean, and smear all bolts with grease. Check that none is bent.

3 When all the bolts are inserted, tighten them fully.

4 On 'F' models, the breather pipe from the rocker cover goes to the breather filter under the air filter. The drain tube from the breather filter passes through a wire clip on the centre stand pivot clamp bolt. On the CB 400F model, the drain tube from the air manifold passes through a lug at the rear of the gearbox.

5 Adjust the clutch to give 10 - 20 mm (7/16 - 7/8 inch) free play at the end of the handlebar lever. At the same time, the clutch should not slip or drag. (If, with the engine running and in gear, the bike should not stall or creep forwards).

6 CB 400F model: Screw the adjusters at both ends of the cable in fully, to give maximum free play in the cable. Remove the triangular clutch adjuster cover, secured by two screws. Slacken the locknut on the clutch adjuster screw and turn the adjuster screw clockwise until resistance is felt (try this several times to get the 'feel'). Now turn the screw back ¼ - ½ turn, and tighten the locknut. Unscrew the gearbox end cable adjuster until there is the required free play at the clutch handlebar

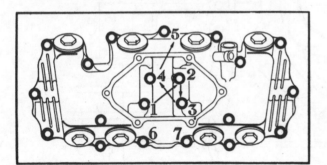

Fig. 1.24. Rocker cover bolt tightening sequence, CB 400F model

Fig. 1.25. Advance the crankshaft 15° to align the spring post (CB 550 models only)

50.8 Replace the cam chain tensioner holder (CB 400F model only)

50.11 Align timing marks as shown

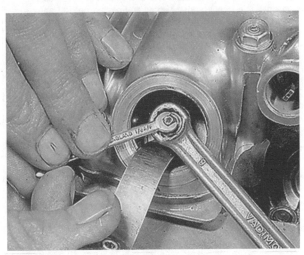

50.12 Hold square on rocker arm when tightening locknut

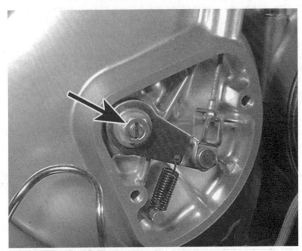

51.6a Coarse adjustment is effected by this screw

51.6b Make fine adjustments here

lever, then tighten the locknut. Fine adjustment may be made at the handlebar end adjuster.

7   Check that the clutch does not slip or drag after every adjustment.

8   CB 550 models: Slacken the adjusters at each end of the clutch cable, to give maximum free play. Slacken the locknut on the clutch adjuster screw and turn the adjuster screw anti clockwise until resistance is felt (try this several times to get the 'feel'). Now screw the adjuster screw in ¼ turn, and tighten the locknut. Unscrew the gearbox end cable adjuster to give the specified play at the end of the handlebar lever, and tighten the locknut. Fine adjustment can be made at the handlebar end adjuster.

9   Reconnect the throttle cables, adjust and check their operation (See Chapter 2).

10 Hook the rear chain over the drive sprocket. Replace the link when fitted. The closed end of the spring link **must** face in the direction of travel of the chain.

11 Replace the exhaust pipe. Always use new exhaust pipe gaskets when replacing the exhaust system. Tighten the exhaust pipe to cylinder head nuts before the silencer to frame fixings. On the CB 400F, the exhaust pipe to silencer manifold clips must be tightened before replacing the exhaust. The bolts in the clips must be above the exhaust pipe.

12 Make sure all electrical connections are tight, and connected correctly. Switch on the ignition and check that the oil pressure warning light and neutral indicator light comes on, and that all the electrical accessories work.
13 CB 400F model: The throttle cable goes through a wire clip by the horn, along with the clutch cable. The latter also passes through a clip on the breather cover.
CB 550 models: The spark plug leads to cylinder 1 and 4 go through clips on the breather cover. The rev-counter cables pass through a clip alongside the horn. The throttle cables are clipped to the top frame spine.
14 Refill the engine with oil after replacing the sump, and turn it over many times to fill the oil galleries, using the kickstart.
15 Think carefully over what has been done and check that all is assembled and tightened correctly.

52 Starting and running the rebuilt engine

1 Turn on the petrol and check that there are no leaks.
2 Start the engine, and keep it running at a low speed for a few minutes to allow oil pressure to build up and the oil to circulate. If the oil pressure warning lamp is not extinguished, stop the engine immediately and investigate the lack of pressure
3 The engine may tend to smoke through the exhausts initially, due to the amount of oil used when assembling the components. The excess of oil should gradually burn away as the engine settles down.
4 Check the exterior of the machine for oil leaks or blowing gaskets. Make sure that each gear engages correctly, and that all the controls function effectively, particularly the brakes. This is an essential last check before taking the machine on the road.
5 CB 400F model only: Retension the cam chain with the engine running. This is accomplished by letting the engine idle at 1200 rpm then loosening the locknut of the adjuster and turning the centre bolt outwards a couple of turns or so. Turn it again inwards, until slight resistance is felt, then re-tighten the locknut. Chain tension is now correct.

53 Taking the rebuilt machine on the road

1 Any rebuilt machine will need time to settle down, even if parts have been replaced in their original order. For this reason it is highly advisable to treat the machine gently for the first few miles to ensure oil has circulated throughout the lubrication system and that any new parts fitted have begun to bed down.
2 Even greater care is necessary if the engine has been rebored or if a new crankshaft has been fitted. In the case of a rebore, the engine will have to be run-in again, as if the machine were new. This means greater use of the gearbox and a restraining hand on the throttle until at least 500 miles have been covered. There is no point in keeping to any set speed limit; the main requirement is to keep a light loading on the engine, and to gradually work up performance until the 500 mile mark is reached. These recommendations can be lessened to an extent when only a new crankshaft is fitted. Experience is the best guide since it is easy to tell when an engine is running freely.
3 If at any time a lubrication failure is suspected, stop the engine immediately and investigate the cause. If an engine is run without oil, even for a short period, irreparable engine damage is inevitable.
4 When the engine has cooled down completely after the initial run, re-check the various settings, especially the valve clearances. During the run most of the engine components will have settled into their normal working locations.

54 Fault diagnosis: engine

Symptom	Cause	Remedy
Engine will not start	Defective spark plugs	Remove the plugs and lay on cylinder heads. Check whether spark occurs when ignition is switched on and engine rotated.
	Dirty or closed contact breaker points	Check condition of points and whether gap is correct.
	Faulty or disconnected condenser	Check whether points arc when separated. Renew condenser if evidence of arcing.
	Valve stuck in guide	Free and clean both stem and valve guide. Renew, if binding.
	Faulty valve timing	Check and re-set.
Engine runs unevenly	Ignition and/or fuel system fault	Check each system independently, as though engine will not start.
	Blowing cylinder head gasket	Leak should be evident from oil leakage where gas escapes.
	Incorrect ignition timing	Check accuracy and if necessary re-set.
	Incorrect tappet clearance	Check and adjust.
Lack of power	Fault in fuel system or incorrect ignition timing	See above.
	Valve sticking	See above.
	Valve seats pitted	Grind in valves
	Weak valve spring	Renew as a set.
	Faulty piston ring	Renew as a set.
Engine overheats	Heavy carbon deposit	Decoke engine.
	Lean fuel mixture	Adjust carburettors.
	Retarded ignition timing	Check and re-set.
Heavy oil consumption	Cylinder block in need of rebore	Check for bore wear, rebore and fit oversize pistons if required.
	Damaged oil seals	Check engine for oil leaks.
	Excessive oil pressure	Check pressure relief valve action.

Excessive mechanical noise	Worn cylinder block (piston slap) Worn camshaft drive chain (rattle) Worn big end bearings (knock) Worn main bearings (rumble)	Rebore and fit oversize pistons. Adjust tensioner or replace chain. Fit replacement crankshaft assembly. Fit new journal bearings and seals. Renew crankshaft assembly if centre bearings are worn.
Engine overheats and fades	Lubrication failure	Stop engine and check whether internal parts are receiving oil. Check oil level in crankcase.

55 Fault diagnosis: clutch

Symptom	Cause	Remedy
Engine speed increases as shown by tachometer but machine does not respond	Clutch slip	Check clutch adjustment for free play at handlebar lever. Check thickness of bonded plates.
Difficulty in engaging gears. Gear changes jerky and machine creeps forward when clutch is withdrawn. Difficulty in selecting neutral	Clutch drag	Check clutch adjustment for too much free play. Check clutch drums for indentations in slots and clutch plates for burrs on tongues. Dress with file if damage not too great.
Clutch operation stiff	Damaged, trapped or frayed control cable	Check cable and renew if necessary. Make sure cable is lubricated and has no sharp bends.

56 Fault diagnosis: gearbox

Symptom	Cause	Remedy
Difficulty in engaging gears	Selector forks bent Gear clusters not assembled correctly	Renew. Check gear cluster arrangement and position of thrust washers.
Machine jumps out of gear	Worn dogs on ends of gear pinions Stopper arms not seating correctly	Renew worn pinions. Remove right-hand crankcase cover and check stopper arm action.
Gearchange lever does not return to original position	Broken return spring	Renew spring.
Kickstart does not return when engine is turned over or started	Broken or poorly tensioned return spring	Renew spring.
Kickstart slips	Ratchet assembly worn	Renew all worn parts.

Chapter 2 Fuel system and lubrication

Contents

Specifications

Carburettors	CB400F	CB550	CB550F
Type		Kei-hin, piston valve type, push-pull operated, with butterfly chokes	
Choke size	20 mm	22 mm	22 mm
Main jet	No. 75	No. 100	No. 98
Pilot jet	No. 40	No. 40	No. 38
Throttle slide	—	2.5	—
Air screw opening ...	2 turns $\pm$ ½	1½ turns $\pm$ 3/8	1¾ turns $\pm$ ½
Jet needle setting ...	Third groove	—	—
Air filter		Renewable paper element	
Carburettor vacuum reading	16 - 24 cm Hg	16 - 24 cm Hg	16 - 24 cm Hg
Engine oil capacity ...	3.5 litres (6.2 Imp. pints) (3.7 US quarts)	3.0 litres (2.6 Imp. quarts) (3.2 US quarts)	3.2 litres (5.3 Imp. pints) (3.4 US quarts)
Oil pump	Double trochoidal gear-type	Trochoidal gear-type	Trochoidal gear-type
Clearance, rotor to housing, max.	0.35 mm (0.0138 in.)	0.35 mm (0.0138 in.)	0.35 mm (0.0138 in.)
Clearance, inner to outer rotor, max.	0.30 mm (0.0118 in.)	0.35 mm (0.0138 in.)	0.35 mm (0.0138 in.)
Oil pressure	4.5 kg/cm^2 (64 lb/in^2)	4.5 kg/cm^2 (64 lb/in^2)	4.5 kg/cm^2 (64 lb/in^2)

Torque wrench settings		kg m	lb ft
Carburettor manifold bolts		7.0 - 1.1	5.1 - 8.0
Oil pressure switch		1.5 - 2.0	10.8 - 14.5
Oil filter centre bolt		2.7 - 3.3	19.5 - 23.8
Sump drain plug		3.5 - 4.0	25.3 - 28.9
Oil gallery plug		1.0 - 1.4	7.2 - 10.1

1 General description

The Honda 400 and 550 fours have four separate carburettors with cylindrical throttle slides and butterfly cold start chokes. The four slides are operated by a push-pull cable, via a shaft mounted above the carburettors. Operating the interconnected chokes on the CB 400F model also raises the throttle slides for starting. The float chamber is concentric with the main jet, and fuel level is controlled by a plastic float.

Fuel is gravity fed from the petrol tank to each float chamber. The single petrol tap with filter, on the left-hand side of the tank, has a reserve position.

Air passing over the needle jet draws petrol from the float chamber, through the main jet. The petrol is atomised in the venturi behind the throttle slide. As the slide rises, the tapered jet needle allows more petrol to emerge from the needle jet. At small throttle openings, a secondary air passage connecting with the pilot jet provides a metered mixture supply. Transition between pilot and main jets is smoothed by the cut-away on the atmospheric side of the throttle slide.

2 Petrol tank: removing and replacing

1 Before removing the petrol tank, turn off the petrol tap and pull off the fuel pipes. Disconnect the tank breather pipe at the junction underneath the tank.

2 Raise the seat, unclip the rubber retainer at the rear of the tank; and lift the tank and pull rearwards to remove.

3 While the tank is removed, check that both sections of the breather pipe are unblocked and free from cracks. A blocked breather pipe will cause fuel starvation. On the CB 400F model the vent hole in the filler cap should be checked, for the same reason.

4 Replace the tank by engaging the semi-circular housings at the front with the rubbers on the frame and clipping the rear strap over the lip at the back of the tank. Make sure that no wires or cables are trapped beneath the tank. Reconnect the petrol pipes.

5 Take care when removing the tank, to avoid fire or explosion. Store it well away from any naked flames.

3 Petrol tap: cleaning the filter

1 CB 550 model: If only the petrol tap filter requires attention, there is no need to remove the tap or to drain the petrol tank. The filter bowl. which has a hexagon head to aid removal, is threaded into the base of the petrol tap and can be unscrewed after the tap has been turned off. The circular filter gauze will also be released and can be washed with petrol to remove any sediment. Before replacing, the filter bowl should be cleaned thoroughly.

2 CB 400F and 550F models: The tap on these models has to be removed to clean the filter, consequently all petrol must first be drained.

3 Unscrew the hexagonal union nut above the tap, and remove the tap carefully with the filter attached. Pull the filter gauze out of the tap and wash in petrol. When replacing the filter, note that the float on its base must locate with the flat on the tap. Don't forget the rubber washer in the union nut. The nut is threaded left-hand for the tap, and right-hand for the tank. Offer the tap up to the tank, and tighten the nut. The plain part of the nut goes above the hexagon.

4 It is seldom necessary to remove the lever which operates the petrol tap, although occasions may occur when a leakage develops at the joint. Although the tank must be drained before the lever assembly can be removed, there is no need to disturb the body of the tap.

5 To dismantle the lever assembly, remove the two crosshead screws passing through the plate on which the operating positions are inscribed. The plate can then be lifted away, followed by a spring, the lever itself, and the seal behind the lever. The seal will have to be renewed if leakage has occurred. Reassemble the tap in the reverse order. Gasket cement or any other sealing medium is NOT necessary to secure a petrol tight seal.

6 CB 550 model only: If the tap body has to be removed, it is held to the underside of the petrol tank by two crosshead screws with washers. Note that there is an O ring seal between the petrol tap body and the petrol tank, which must be renewed if it is damaged or if petrol leakage has occurred.

7 When draining the petrol tank, **DO NOT SMOKE.** There is danger of explosion or fire.

4 Petrol feed pipes: examination

1 Synthetic rubber feed pipes are used to convey petrol from the petrol tap to the float chamber of each of the four carburettors. Each pipe is retained by a wire clip, which must hold the pipe firmly in position. Check periodically to ensure the pipes have not begun to split or crack and that the wire clips have not worn through.

2 Do NOT replace a broken pipe with one of natural rubber, even temporarily. Petrol causes natural rubber to swell very rapidly and disintegrate, with the result that minute particles of rubber would easily pass into the carburettors and block the internal passageways. Plastic pipe of the correct bore can be used as a temporary substitute, but it should be replaced with the correct type of tubing as soon as possible since it will not have the same degree of flexibility.

5 Carburettors: removing

1 Lift the seat, and take out the tool tray and the air filter top. Extract the air filter by pulling on the V-shaped clip. Remove the petrol tank.

2 Slacken the air manifold to carburettor clips, and on the 400 the manifold to air cleaner box clip. Also on the CB 400F model, prize out the oval metal sleeve between the manifold and the air filter box, from inside the filter box.

3 CB 550 model only: Remove the two filter box to manifold

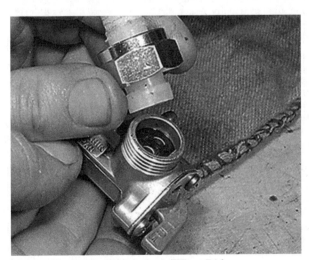

3.3 Align the flats on filter and tap ('F' models)

5.5 Disconnect the throttle cables

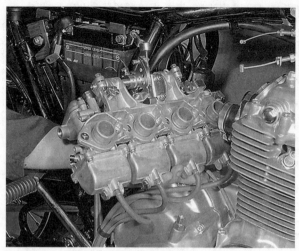

5.6 Remove the carburettor assembly

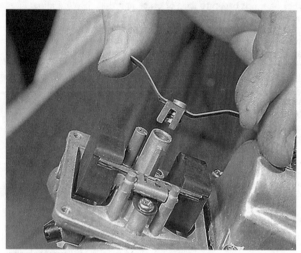

6.3 Pull off the leaf spring and main jet

6.4 Unscrew the pilot jet

bolts, one accessible from inside the filter box, the other from outside at the bottom;

4   Pull off the drain tube from the bottom of the air manifold, when fitted. Pull the manifold off of the carburettor stubs, and wriggle it clear.

5   Slacken the two throttle cable locknuts, remove the adjusters from their housings, and unhook the nipples from the drum.

6   Slacken the four carburettor to cylinder head clips, and pull the carburettor assembly off of the induction stubs. The drain and breather tubes should be pulled out from behind the gearbox, at the same time.

6   Carburettors: dismantling, cleaning and reassembling

1   The four carburettors are bolted to one casting, which also forms the bracket for the operating mechanism. It will not be necessary to remove the carburettors from this casting, unless a complete item has to be renewed.

2   Dismantle each carburettor in turn, so as not to mix up parts. Remove the float bowl after unscrewing the four crosshead fixing screws.

3   Pull out the leaf spring jet clip and main jet together. Tap the carburettor to remove the needle jet (see illustration). Make sure each jet is clear before replacing it. Examine the needle jet for wear, it will wear oval after lengthy service. Renew both needle and jet together. Examine the O ring seal on the main jet. The slot in the leaf spring engages with the ridge on the jet holder.

4   Unscrew the pilot jet, and check that it is clear before replacing it.

5   Extract the float pivot pin to remove the floats. Shake the hollow type plastic floats to check if any petrol is inside, which will indicate a puncture. Solid floats will not, of course, puncture.

6   Unscrew the needle valve retainer, and pull out the valve. Examine the float needle and its seating in the valve, for wear. Renew the pair if there is a ridge worn in the conical tip of the needle. Examine the O ring seal on the valve. The needle valve drops out very easily; take care not to lose it.

7   Replace the needle valve, and the float and pivot pin. To ensure correct fuel level, check the float height as follows: With the float arm only just touching the needle jet, measure the height of the top of the float from the float bowl mounting face. Adjust by bending the float arm. The float height should be 21 mm (0.83 in) for 400 models, 14.5 mm (0.57 in) for 1977 CB550K3 models, and 22 mm (0.87 in) for all other 550 models.

8   Replace the float bowl. Ensure that the gasket is in good condition, and seated correctly. Overtightening the float bowl fixing screws will only distort it, and increase any leak. Leaks are caused by a faulty gasket, dirt or a previously distorted float bowl.

9   Never use a piece of wire or any pointed metal object to clear a blocked jet. It is only too easy to enlarge the jet under these circumstances, and increase the rate of petrol consumption. If compressed air is not available, a blast of air from a tyre pump will usually suffice.

10 Unhook the throttle return spring between the centre carburettors. Unscrew the two carburettor top fixing screws from each carburettor and remove the top. Position the throttle valve to full open and straighten the tab washers of the two hexagon headed bolts. Remove the 4 mm bolt from the shaft end and loosen the 6 mm bolt on the throttle side about half a turn. Insert a screwdriver between the throttle shaft and link arm and pry the link arm free.

11 Withdraw the throttle valve and needle assembly from the carburettor body. Unscrew the two 3 mm screws and remove the valve plate and the jet needle from the throttle valve. Examine the throttle needle for wear and renew, together with the needle jet if necessary. Fit the needle clip in the correct groove.

12 The manually-operated chokes are unlikely to require attention throughout the normal service life of the machine. When the operating plungers are depressed, flaps are lowered into the carburettor air intake which cut off the supply of air and therefore give a much richer mixture for cold starting. The

61

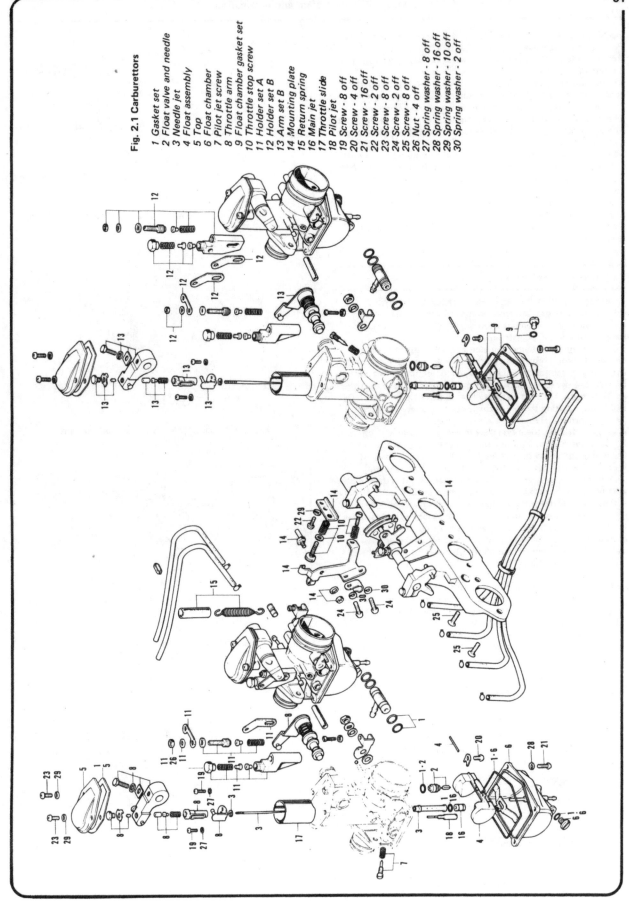

Fig. 2.1 Carburettors

1 Gasket set
2 Float valve and needle
3 Needle jet
4 Float assembly
5 Top
6 Float chamber
7 Pilot jet screw
8 Throttle arm
9 Float chamber gasket set
10 Throttle stop screw
11 Holder set A
12 Holder set B
13 Arm set B
14 Mounting plate
15 Return spring
16 Main jet
17 Throttle slide
18 Pilot jet
19 Screw - 8 off
20 Screw - 4 off
21 Screw - 16 off
22 Screw - 2 off
23 Screw - 8 off
24 Screw - 2 off
25 Screw - 8 off
26 Nut - 4 off
27 Spring washer - 8 off
28 Spring washer - 16 off
29 Spring washer - 10 off
30 Spring washer - 2 off

machine should never be run for any distance with the chokes
closed or the excessively rich mixture will foul the spark plugs
and wash the oil from the cylinder walls, greatly accelerating
the rate of engine wear.

13 Do not use excessive force when reassembling a carburettor
because it is easy to shear a jet or some of the smaller screws.
Furthermore, the carburettors are cast in a zinc-based alloy
which itself does not have a high tensile strength. Take particular
care when replacing the throttle slide to ensure the needle aligns
with the jet seat.

14 Do NOT turn either the throttle stop or pilot air screw with
out noting their exact positions. Failure to do so will necessitate
re-synchronising the carburettors.

15 If a carburettor or carburettors have been removed from the
mounting casting, be sure to reconnect all of the interconnecting
tubes and 'O' ring seals etc. On the CB 550 models, fuel pipe
feeds each pair of carburettors. The CB 400F model has a single
feed pipe for all four carburettors.

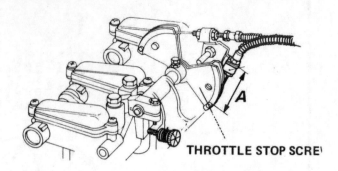

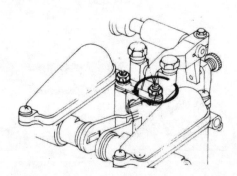

Fig. 2.3 Distance 'A' must be adjusted before synchronising
carburettors

7 Carburettors: adjusting

1 Before adjusting the carburettors, check the following:
Contact breaker gap and ignition timing; valve clearances and
spark plug gaps. The engine should be at its normal working
temperature during adjustment.

2 Do not run the engine in a confined space, for there is a
danger of carbon monoxide poisoning from the exhaust gases.
If the engine cannot be run outside, good ventilation is essential.

3 Remove the air manifold, and check that when the throttles
are fully open, the slide to bore distance is 0 - 1.0 mm (0 - 0.04
inch). Adjust using the full throttle opening screw, if necessary
(see illustration). Replace the air manifold.

4 Adjust the throttle stop screw to give a tickover at 1200 rpm.

5 Carburettor synchronising must be checked with a set of
vacuum gauges. Few owners are likely to own such a set, which
is expensive, and normally held by Honda agents, who will
synchronise the carburettors for a nominal sum. It is questionable
whether there is any advantage in buying a set of gauges. However
if they are available, instructions for synchronising are given.

6 First, check the vacuum readings: The gauge adaptors screw
into tapped holes in the inlet ports, each of which is normally
plugged with a crosshead screw. The reading on the gauges
should be 16 - 24 cm Hg and all the gauges must read within
3 cm Hg of each other.

7 If adjustment is called for, it is necessary to remove the petrol
tank, and mount it about 50 cm (20 inch) higher, with longer
fuel pipes. Adjust the throttle stop screw until the distance 'A'
between the end of the bottom cable, and the cable nipple, is
56 mm (2.2 inch) (See illustration).

8 Start the engine, and note the readings on the gauges. Slacken
the locknut, and turn the adjuster on the throttle linkage of the
carburettor requiring adjustment (see illustration). Tighten the
locknut. Snap the throttle open and closed several times, and
re-check the gauges. Adjust each pilot air screw to give maximum
vacuum reading and good tickover. The screw should be open

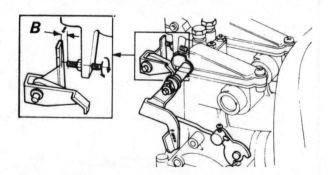

Fig. 2.4 Carburettor synchronising adjustment

Fig. 2.5 Fast idle speed adjuster (400 only)
Adjust clearance 'B'

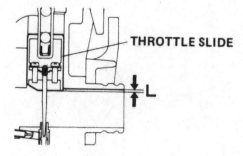

Fig. 2.2 Measuring throttle slide to bore distance

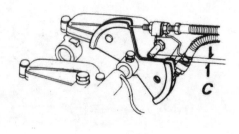

Fig. 2.6 Overtravel stop adjuster
Adjust clearance 'C'

6.5 Extract the float pivot pin

6.6 Pull out the float needle valve

6.7 Measure from top of float to float bowl mounting face

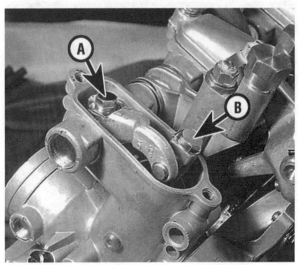

6.10 A: 4mm shaft end bolt B: 6mm throttle side bolt

7.4 Throttle stop screw

7.6 Vacuum gauge hole plugs

7.8 Pilot air screw

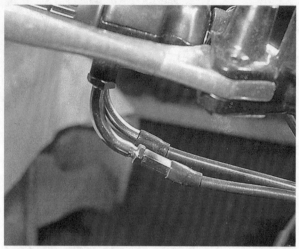

7.12 Throttle cable adjuster

8.1 Pull out air filter element

8.3 Squeeze end of drain tube to clean it

Fig. 2.7 Engine breather filter
components (CB 400F model)

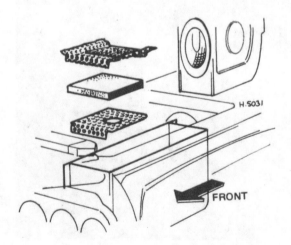

1½ ± ½ turns on the CB 550 models and 2 ± ½ turn on the CB 400F model. Remove the vacuum gauge adaptors, and replug the adaptor holes.

9 If needle swing on a vacuum gauge is too great, tighten the gauge restrictor valve so that swing is limited to within 2 cm Hg. If the indicated pressure is lower than 15 cm Hg, suspect an air leak or absence of slack in the throttle cable. Always pause after making an adjustment to the pilot jet screw to allow engine rpm to stabilise.

10 CB 400F model only: Fast idle speed should be adjusted after synchronising carburettors. Check that, when the cold start choke is fully open, the clearance 'B' between the stop and the adjusting screw is 0 - 0.3 mm (0 - 0.012 inch) (see illustration). Loosen the locknut and adjust the screw, if necessary.

11 Overtravel stop: Close the throttle completely. The clearance 'C' between the stop pin and the cable drum should be 2 - 2.1 mm (0.079 - 0.083 inch). (See illustration). To adjust, slacken the locknut and turn the eccentric pin.

12 Adjust the throttle cables to give minimum play at the twist grip. Slacken the locknut on the adjuster at the carburettor end, and turn the adjuster to give the required free-play. Fine adjustment may be made using the adjuster at the twist grip end.

13 Carburettor jet sizes etc, are selected by the manufacturer after exhaustive tests, using the recommended fuel grade. They should not need to be altered, other than in exceptional circumstances.

8 Air filter - cleaning and renewing

1 Service the air cleaner at the recommended intervals, or sooner in dusty conditions. A clogged filter will reduce performance and increase fuel consumption. Lift the seat and remove the tool tray and air filter cover. Pull on the U-shaped clip, and extract the filter element. Also check that the hole in the bottom of the air filter case is not blocked (CB 550 model only).

2 On the 'F' models, an engine breather filter is incorporated. This filters oil from the engine blow-by gases, and re-cycles them through the carburettors. After removing the air filter, take out the perforated plate and the breather element in the bottom of the filter case. Wash the breather element in solvent (other than petrol), squeeze out the excess solvent and dry the element. Do not wash the element in acid, alkali or organic solvent.

3 A drain tube leads from the bottom of the air filter case, the end of which should be squeezed to clear accumulated oil or water.

9 Exhaust system: examination and renovation

1 Removal of the exhaust pipes and silencers is described in Chapter 1, Section 4.

2 Do not alter any part of the exhaust system, it has been designed by the manufacturers after exhaustive testing. The exhaust pipe gaskets must always be fitted. Leaks from this joint will cause weak running, and possibly a holed piston.

3 On the CB 400F model, two of the exhaust pipes may be detached from the silencer manifold. Check the condition of the exhaust pipe to manifold sealing gaskets. The bolts of the exhaust pipe clips must be positioned above the pipes, and tightened before refitting the assembly to the machine.

4 The exhaust pipe and silencer on the CB 550 model are in one piece, and have to be replaced in one. The silencers are connected at the ends by a rubber bush with a clip. Do not omit this bush.

5 On the CB 550F model, the single silencer may be removed from the exhaust pipes. The joint cover has to be removed first, it is retained by a bolt and two straps. The straps hook onto ears on the cover. Slacken the silencer clamp bolt, and pull off the silencer. Check the silencer sealing gasket before replacing.

10 Engine lubrication: general description

Lubrication is by wet sump system, with the gearbox sharing the engine oil. The trochoidal gear pump is driven by the gearbox input shaft. Oil is drawn from the sump through a wire mesh strainer, and pumped through a paper element filter, to a main oil gallery running longitudinally beneath the crankshaft. This gallery feeds all five main bearing journals, and the big-ends. Oil from the two outer journals is fed, via restrictors, to the camshaft bearings; and to oil jets on the 400 model only, which spray the cam flanks.

The rockers, valve stems and cam chain, are lubricated by splash. Oil drains down the cam chain tunnel to the sump

On the CB 400F model, all gear shafts and the clutch are pressure lubricated from a smaller rotor in tandem with the main oil pump.

The primary chain dips into the sump oil.

There is a pressure relief valve on the oil pump, to prevent damage and a low oil pressure warning lamp.

11 Engine oil: checking level, changing oil and filter

1 Engine oil level should be checked regularly, preferably each time the bike is used. The machine should be upright on a level surface. Do not screw in the dipstick, just rest it on the boss. Keep oil level up to the upper mark on the dipstick, but do not overfill. Never run the engine if oil level is below the lower mark.

2 Renew the engine oil at the recommended intervals, and so reduce the mechanical wear on engine components. In ardous or dusty conditions, the oil should be renewed more frequently. The engine should be at normal running temperature when the oil is changed. Put the bike on its centre stand, and a container under the sump - minimum capacity 4 litres (7 pints).

3 Remove the sump drain plug and washer, also unscrew the central filter housing bolt and remove the filter. The filter housing is full of oil too. Allow all the oil to drain out, turn the engine with the kickstart to assist this. Replace the drain plug and filter after cleaning them.

4 Refill with oil, run the engine for a few minutes, and check the oil level on the dipstick.

5 If a filter change is due, do this at the same time. Extract the old filter from the housing. Clean the filter housing inside. Make sure that the spring and the washer are in place at the bottom of the housing, and fit the new filter. Check the condition of the 'O' ring seals. When replacing the filter housing, note that a boss on the crankcase, just below the crankcase joint, locates between two fins of the housing.

12 Oil strainer: removing and cleaning

1 Drain the engine oil when the engine is hot.

2 Remove the sump after unscrewing ten (CB 550 models) or eleven (CB 400F model) sump bolts. Take care not to damage the gasket or sealing ring.

3 Pull off the strainer body under the sump.

4 Extract the filter screen from the strainer, wash in petrol, and replace. Clean out the sump. On the CB 400F model, examine the O ring seal on the strainer spigot.

5 Replace the strainer, connecting up the oil pipe to the spigot under the crankcase. CB 400F model only: don't forget the spring over the strainer spigot.

6 Replace the sump, with a new gasket or sealing ring if necessary. Tighten the drain plug and refill with oil.

7 There are three (CB 400 model) or four (CB 550 models) main oil gallery plugs. One of these is pushed in, with an O ring seal, and is retained by the alternator cover. The others are external and are screwed in; also having 'O' ring seals. One plug is below the contact breaker housing, the other alongside the oil filter. Check the security, and the condition of the seals on all

11.3a Sump drain plug (CB 400F)

11.3b Remove the filter and housing

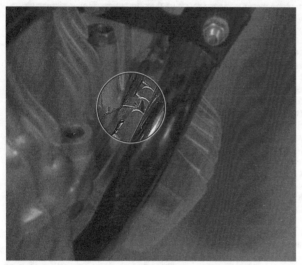

11.5 Fins on filter housing locate with a boss (CB 400F)

12.3 Pull off the strainer body

12.5 The oil pipe from strainer pushes onto a spigot

13.6 Remove the oil pump (CB 400F illustrated)

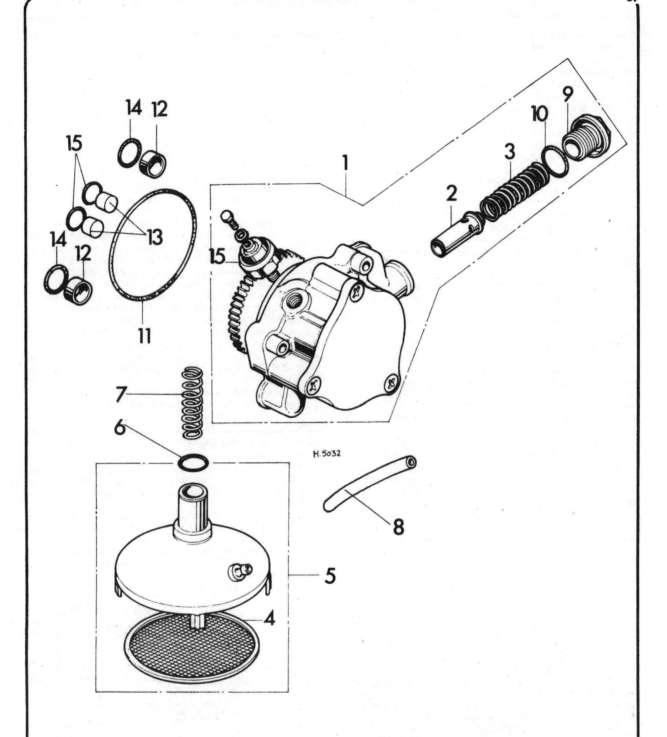

H.5032

Fig. 2.8 Oil pump and strainer

1 Oil pump complete
2 Relief valve plunger
3 Relief valve spring
4 Oil filter screen
5 Oil strainer body

6 'O' ring seal
   (CB 400F model only)
7 Spring
   (CB 400F model only)
8 Oil pipe

9 Relief valve cap
   (CB 550 models only)
10 'O' ring seal
   (CB 550 models only)
11 'O' ring seal
12 Dowel - 2 off

13 Dowel - 2 off
   (CB 400F model only)
14 'O' ring seal - 2 off
15 'O' ring seal - 2 off
   (CB 400F model only)
16 Oil pressure switch

plugs. By removing a screw-in plug, the supply of oil to the main bearings can be verified.

13 Oil pump: removing, dismantling and reassembly

1   The oil pump may be removed with the engine in frame. However, do not dismantle the pump unnecessarily, only if wear or pump failure are suspected. Reassembly must be done in clinically clean conditions.
2   Drain the engine oil when the engine is hot.
3   CB 400F model: remove the gear lever circlip and washer. Remove the gear lever pinch bolt, and pull off the gearchange assembly. Replace the pinch bolt. CB 550 models: Remove the starter motor cover after unscrewing the two hexagonal bolts. Remove the gear pedal pinch bolt, and pull off the gearchange assembly. Replace the pinch bolt.
4   Unscrew the four crosshead screws (an impact driver may be needed), and take off the gearbox cover on the left-hand side.
5   Disconnect the oil pressure switch wire.
6   CB 400F model: Unscrew the three small hexagonal head bolts and one larger bolt and pull the pump out of its housing (note the cable clip on the front top bolt). CB 550 models: Unscrew the three crosshead screws which secure the pump, and pull it out of the housing.
7   CB 400F model: Unscrew the two crosshead screws at the drive end of the pump, and remove the end cover with rotors. Take out the pressure relief valve spring and plunger from the main casting. Extract the inner and outer rotors, noting which way up they were fitted. The drive pinion shaft is retained in the end cover by the rotor drive dowel. Push out the dowel and pull out the shaft. The gearshaft pump should not need attention; the construction is the same.
8   CB 550 models: Remove the three countersunk screws in the end cover of the pump, and take off the end cover. Extract the inner and outer rotors, noting which way up they were fitted. Take out the dowel pin which passes through the rotor shaft, so that the pinion and shaft can be withdrawn, from the other end of the pump body.
   The main oil pump casing contains the oil pressure release valve which is located behind the hexagon headed end cap. The pressure release valve takes the form of a plunger with a shouldered end, loaded by a heavy gauge coil spring.
9   Wash all the oil pump components with petrol and allow them to dry before carrying out a full examination. Before part reassembling the pump for the various measurements to be

made, check the castings for cracks or other damage, especially the pump end cover.
10 The centre pop marks on the rotor flanks must be aligned. Reassemble the pump rotors and measure the clearance between the outer rotor and the pump body, using a feeler gauge. If the clearance exceeds 0.013 inch (0.35 mm) the rotor or the body must be renewed whichever is worn. Measure the clearance between the outer rotor and the inner rotor with a feeler gauge. If this clearance is greater than 0.012 inch (0.30 mm) the rotors must be renewed as a set.
11 Examine the rotors and the pump body for signs of scoring, chipping or other surface damage which will occur if metallic particles find their way into the oil pump assembly. Renewal of the affected parts is the only remedy under these circumstances, bearing in mind that rotors must always be replaced as a matched set.
12 Reassemble the pump and the pump casting by reversing the dismantling procedure. Make sure all parts of the pump are well lubricated before the end cover is replaced and that there is plenty of oil between the inner and outer rotors. Tighten the end cover down evenly and continually check that the drive pinion revolves freely up to the point where the crosshead screws have been tightened fully. A stiff pump is usually due to dirt in the pump assembly. CB 400F only: The flat on the end of the drive spindle must engage the gearbox pump rotor.
13 Inspect, and do not omit the 'O' ring seals. There is a large seal around the pump spigot, and a seal on each of the oilway dowels (two on the CB 550 models, four on the CB 400F models). On the CB 550 models, check that the pressure relief valve cap is tightened fully.
14 Replace the components in the reverse order of dismantling. Tighten the sump drain plug and refill with oil.

14 Oil pressure warning light

1   The oil pressure warning light switch is screwed into the top of the oil pump. It is set to operate at 0.3 kg/sq. cm (4.3 lb/sq in) on the CB 400F model, or 0.5 kg/sq. cm) (7 lb/sq. in) on the CB 550 models).
2   Normal oil presusre is as follows: CB 400F model: 4.5 kg/sq. cm (64 lb/sq. in) CB 550 models: 4 - 6 kg/sq. cm (57 - 85 lb/sq. in).
3   If the oil pressure warning lamp comes on while the motorcycle is being ridden, switch off **immediately** and investigate the cause.
4   Check engine oil level in the sump and top up if necessary.

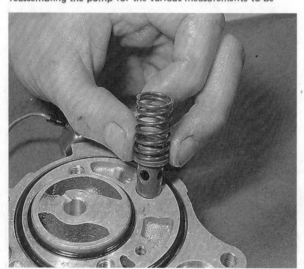

13.7a Take out the oil pressure relief valve (CB 400F model shown)

13.7b Push out the rotor drive dowel ...

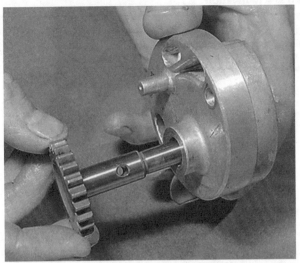

13.7c ... and pull out the shaft (CB 400F model)

13.10a Align the dots on the rotors

13.10b Measure the clearance, outer rotor to housing ...

13.10c ... and that of inner to outer rotor

13.13a There are O-ring seals on the internal and external spigots ...

13.13b ... also on the oilway dowels (CB 400F shown)

Make sure the oil filter strainer and oil passageways are not blocked.

5   If a lubrication system fault is not apparent, check the wiring as described in Chapter 6.

6   The switch may be removed after disconnecting the wire, using a spanner on the large hexagonal portion. When re-tightening the switch, take care not to apply too much force, and shear the mounting studs.

15 Fault diagnosis: fuel system and lubrication

Symptom	Cause	Remedy
Engine gradually fades and stops	Fuel starvation	Check vent hole in filler cap, or breather tube. Sediment in filter bowl or float chamber. Dismantle and clean.
Engine runs badly. Black smoke from exhausts	Carburettor flooding	Dismantle and clean carburettor. Check for punctured float or sticking float needle. Check fuel level in float chamber.
	Air filter blocked	Remove and clean.
Engine lacks response and overheats	Weak mixture Air cleaner disconnected or hose split Modified silencer has upset carburation	Check for partial block in carburettors. Reconnect or renew hose. Replace with original design.
Oil pressure warning light comes on	Lubrication system failure	Stop engine immediately. Trace and rectify fault before re-starting.
Engine gets noisy	Failure to change engine oil when recommended	Drain off old oil and refill with new oil of correct grade. Renew oil filter element.

Chapter 3 Ignition system

Contents

Specifications

	CB400	CB550	CB550F
Ignition		Battery and ignition coil	
Spark plugs - NGK ...	D8ES-L	D7ES	D7ES
ND ...	X-24ES	X-22ES	X-22ES
Spark plug gap	0.7 - 0.8 mm	0.6 - 0.7 mm	0.6 - 0.7 mm
	(0.028 - 0.032 in.)	(0.024 - 0.028 in.)	(0.024 - 0.028 in.)
Contact breaker gap ...	0.3 - 0.4 mm	0.3 - 0.4 mm	0.3 - 0.4 mm
	(0.012 - 0.016 in.)	(0.012 - 0.016 in.)	(0.012 - 0.016 in.)
Capacitor	0.22 microfarads ± 10%	0.22 microfarads ± 10%	0.22 microfarads ± 10%
Ignition timing (full advance)	23.5° - 26.5°	23.5° - 26.5°	23.5° - 26.5°
Advance commences ...	1400 - 1600 rpm	1500 rpm	1500 rpm
Full advance	2300 - 2500 rpm	2500 rpm	2500 rpm

Torque wrench settings	kg m	lb ft
Contact breaker cam bolt	0.8 - 1.2	5.7 - 8.6
Spark plug	1.2 - 1.6	8.6 - 11.6

1 General description

The spark necessary to ignite the petrol/air mixture is generated by two ignition coils, one for each pair of cylinders. Dual contact breakers (180° out of phase), operated by a cam driven from the crankshaft, determine the moment at which the spark will occur. There is an automatic centrifugal spark advance device.

When the contact points open, the low tension circuit to the coils is interrupted, and the magnetic field around the low tension coil collapses. This induces a very high voltage in the high tension coil. The only path for this voltage is across the plug gap, where the resulting spark ignites the mixture. A capacitor connected in parallel with each set of points reduces sparking across them, and so increases their life.

The firing sequence is 1, 2, 4, 3 at each 180° of crankshaft rotation. Hence there is an idle spark in each cycle. The ignition coils are mounted beneath the petrol tank, on each side of the frame spine.

2 Contact breakers - adjustment

1 The contact breakers are located behind the crankcase end cover on the right. This is retained by two cross-head screws, and has a cork gasket which should be removed carefully.

2 Before adjusting the points, examine each set for burning or pitting. Clean or renew the points as necessary. See Section 2. The points are marked '1.4' and '2.3' adjacent to the relevant contact set, indicating the pair of cylinders they serve.

3 Set the gap of points 1.4 first. Turn the crankshaft using a spanner on the large hexagonal washer securing the cam, until points 1.4 are fully open. Measure the gap with a feeler gauge, and adjust if necessary. Standard gap: 0.3 - 0.4 mm (0.012 - 0.16 inch).

If the gap requires adjustment, slacken slightly the slotted screw which secures the fixed contact. A screwdriver should be engaged between the slot in the fixed contact, and the two pins on the contact breaker back plate; by turning the screwdriver the gap may be opened or closed. Tighten the screw and re-check the gap.

4 Turn the crankshaft so that points 2.3 are fully open, and repeat the procedure above. Do not slacken the two screws which secure the 2.3 contact set base plate to the main base plate, this will upset the timing. It is important that both points should be set to the same gap, as the gap determines the moment when the contacts open, and thus the ignition timing.

5 Before replacing the cover and gasket, oil the contact breaker cam lubricating felt. Do not put too much oil on the

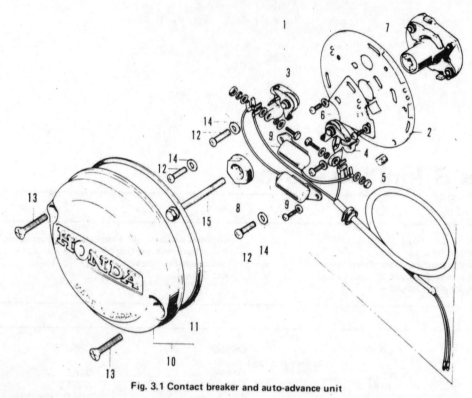

Fig. 3.1 Contact breaker and auto-advance unit

1 Contact breaker assembly	4 2 & 3 cylinder contact breaker	8 Advance shaft washer	12 Screw - 3 off
2 Contact breaker back plate		9 Capacitor - 2 off	13 Screw - 2 off
3 1 & 4 cylinder contact breaker	5 Lubricating felt	10 Points cover	14 Plain washer - 3 off
	6 Adjustable base plate	11 Points cover gasket	15 Bolt
	7 Auto-advance assembly		

felt, if any gets onto the contacts it will cause misfiring.

3 Contact breaker points: examination and renovation

1 Unscrew the retaining screws of the crankcase end cover on the right. Remove the cover and gasket to expose the contact breaker. If, after examination, the points appear to be burned or blackened, they should be removed for renewal or cleaning. An even grey colour is normal.

2 Remove the nut, washers and screws which secure the leads to the contact set terminal. Note carefully the order in which the insulating washers are assembled; if they are replaced incorrectly, there may be a short circuit. Unscrew the slotted screw securing the fixed contact to the base plate, remove the contact pivot, and pull off the moving contact.

3 If the points are only slightly pitted it is possible to dress them with a fine oilstone. The points must be kept absolutely flat and square. If the pair close at an angle, they will quickly be burned away. Badly pitted or worn points must be renewed. Pitting prevents the correct gap being set and performance will suffer.

4 Before replacing the moving contact, smear the pivot lightly with grease. Replace the contact breaker set, and assemble the insulating washers, capacitor lead and contact breaker lead in correct sequence. Check and re-adjust the contact breaker gap.

5 If the points burn away rapidly, suspect the capacitor. See following Section.

4 Capacitors: examination and renewal

1 A capacitor is placed in electrical parallel with each set of contact points to reduce arcing, and thus increase life expectancy. A persistant misfire, or poor starting may be caused by a faulty capacitor. This may occur only when the capacitor has reached working temperature.

2 Since each set of points is common to a pair of cylinders, a misfire on one cylinder only will not be due to a fault capacitor. Similarly, misfiring on all cylinders is unlikely to be due to faulty capacitors, as the chances of both failing together are remote. An exception could occur if the capacitors have been dented or punctured. This could happen in an accident, or if the bike is dropped.

3 A quick check of the capacitor may be made by flicking open the closed points with the ignition switched on. There should be only a faint spark across the points.

4 The capacitors are mounted below the contact breakers. Check that the fixing screws are secure, and good electrical contact exists between the cases and the contact breaker back plate. Similarly, the capacitor lead terminals must be in electrical contact with the moving contact breaker points. Check this point if the contact breaker assembly has been removed.

5 It is not possible to check a capacitor without special equipment. Exchange the leads (blue and yellow) of the capacitors, and run the engine to see if the misfire is transfered to the other pair of cylinders. If a capacitor is proved defective, it will have to be renewed. A damaged capacitor cannot be repaired.

5 Ignition timing: checking and adjusting

1 There are two methods of adjusting the timing, statically and dynamically. If a good quality stroboscope is used correctly, the second method is most accurate. But accurate timing may be obtained if great care is taken with static timing. To check or re-time the ignition statically, a 12V test lamp or buzzer is needed. A buzzer is more convenient, since it is one less item to watch.

2   Remove the crankcase end cover on the right-hand side of the machine. First check the points gap and condition, and service as necessary. Check points 1.4 first. Disconnect the capacitor (blue) lead, and connect the test device between the contact breaker terminal and chassis earth. Turn the ignition on.

3   Turn the crankshaft clockwise with a spanner on the large hexagonal cam fixing washer. The test lamp should light or the buzzer sound, when the 'F 1.4' mark aligns with the static timing mark, viewed through the window in the contact breaker back plate. If it does not do so, and the points gap is correct, adjust the contact breaker opening point.

4   Turn the crankshaft clockwise until the 'F 1.4' mark aligns exactly with the static timing mark. Slacken slightly the three cross-head screws around the periphery of the contact breaker back plate. Turn the whole back plate carefully until the lamp lights (buzzer sounds). Tighten the screws, and recheck the timing. Reconnect the capacitor lead. Turning the back plate clockwise retards the ignition, and vice versa.

5   Check points 2.3 in the same way, disconnecting the yellow capacitor lead, and aligning the 'F 2.3' mark. If points 1.4 required adjustment, it is inevitable that 2.3 will require it too. Points set 2.3 is fixed to a secondary back plate **which can be adjusted independantly of the main back plate.** Turn the crankshaft clockwise until the 'F 2.3' mark aligns exactly with the static timing mark. Slacken slightly the two screws securing the secondary back plate. Insert a screwdriver between the two pins on the main back plate, and the slot in the seconoary plate. Turn the screwdriver to to move the points set until the lamp lights (or buzzer sounds). Tighten the screws and recheck the timing. Reconnect the capacitor lead.

6   If it is not possible to time the ignition within the adjustment provided, check that the automatic advance unit is not jammed; see following Section.

7   Optimum performance depends on the accuracy with which the timing is set. Even a small error can cause a marked reduction in performance, and in the extreme, engine damage through overheating. As already mentioned, this accuracy can be best achieved using a good quality stroboscope, which has a high light output to make using it easier. When timing the ignition dynamically, do not confuse the 'F' timing mark with the adjacent 'T' top dead centre mark. Turning the crankshaft further anti clockwise will expose the two full advance timing marks (these have no other markings) close together.

8   Time cylinders 1 and 4 first. Connect the strobe in accordance with the makers instructions. Check the condition of the points, and overhaul as necessary. Start the engine and aim the light at the timing window. At idling speed (1200 rpm for 400 and 1000 rpm for 550's) the 'F 1.4' mark should align exactly with the static timing mark. Increase engine speed to 2500 rpm. The timing should now have advanced about 25°; so that the two advance marks are on either side of the static mark (they may not be exact). Adjust the back plate if necessary by slackening slightly the three cross head screws round the periphery of the plate and turning it clockwise to retard or anticlockwise to advance. Tighten the screws and recheck. If the timing marks do not advance as engine speed is increased, or if there is insufficient adjustment, check the automatic advance unit.

9   Check cylinders 2 and 3 in the same way, aligning 'F 2.3' marks with the static timing mark. Adjust if necessary as described in paragraph 5.

10 The contact breaker gaps must be checked and set correctly **before** setting the ignition timing. Adjustment of the gaps after wards will alter the timing.

---

6   Automatic advance unit: examination and renovation

---

1   The automatic advance unit rarely requires attention, although it is advisable to examine it periodically, when the contact breaker is overhauled. Most problems arise as a result of condensation within the engine which causes rust to seize, or restrict the

2.3 Check contact breaker gap with a feeler gauge

4.4 The capacitors are mounted below the contact breakers

5.3 The 'F' timing mark aligned with the static mark ...

5.7 ... and the two advance marks

6.3 Unscrew the cam fixing bolt ...

6.4 ... and the three backplate screws

movement of the bobweights.

2   The mechanism has two spring loaded bobweights, which are thrown outwards by centrifugal force. At idling speed the force cannot overcome spring pressure. Above idling speed the force is sufficient to throw the bobweights to the limit of their travel, ie; full advance.

3   Remove the end cover on the right-hand end of the crankcase. Unscrew the centre cam fixing bolt with its large hexagonal washer. Mark with a centre punch the contact breaker back plate and the housing to aid replacement in the same position.

4   Unscrew the three screws around the periphery of the back plate and remove the back plate. It may be left attached to its leads, resting securely on the gearbox. Pull the advance unit off the crankshaft, taking care not to drop the drive dowel.

5   Check that the bobweights move freely, and that their pivots are unworn and rust free. Check spring tension: 680 - 850 grammes (1.43 - 1.87 lb). Lubricate the pivots.

6   Replace the advance unit, locating the drive dowel in the hole in the crankshaft. Replace all other components in reverse order. Note that the large hexagonal washer fits over two pegs on the end of the cam. Check the points gap and the ignition timing.

7   If the automatic advance unit sticks in retard, the engine will overheat, and possibly be damaged. If the unit sticks in the full advance position, starting and idling will be poor.

7   Ignition coils: examination

1   Each ignition coil is a sealed unit, designed to give long service without need for attention. Each coil serves a pair of cylinders. The coils are located beneath the petrol tank, fixed to the frame spine.

2   If a weak spark and difficult starting causes the performance of a coil to be suspect, it should be tested by a Honda agent or an auto-electrical engineer who will have the appropriate test equipment. A faulty coil must be renewed; it is not possible to effect a satisfactory repair.

3   A defective capacitor in the contact breaker circuit can give the illusion of a defective coil and for this reason it is advisable to investigate the condition of the capacitor before condemning the ignition coil. Refer to Section 3 of this Chapter for the appropriate details.

4   Note that it is unlikely for both ignition coils to develop a fault at the same time, unless the common electrical wiring is faulty. Similarly, misfiring on all cylinders is unlikely to be due to both coils being faulty.

8   Spark plugs: examination

1   A matched set of 12 mm medium reach plugs is fitted to the Honda CB 400F and CB 550 fours. The recommended grades and equivalents are given in the Specification. Do not deviate from those recommended; they will give the best all-round results.

2   The plug gap should be set to the dimension given in the Specification. (Most new plugs are supplied with this gap). Check the gap with a feeler gauge and adjust if necessary by bending the outer electrode, never the centre one. If the electrodes are very worn, renew the set of plugs.

3   Clean the electrodes with a wire brush and wipe the porcelain insulator. Check that the insulator is not broken or cracked, especially around the centre electrode. Check that the plug threads are clean, and the copper washer is in position. Tighten the plug only sufficiently to make a gas-tight seal, overtightening may strip the cylinder head threads.

4   With some experience, the condition of the spark plug electrodes and insulator can be used as a reliable guide to engine operating conditions. See the accompanying photographs.

5   Always carry a spare set of spark plugs of the recommended grade. In the rare event of plug failure, they will enable the engine to be restarted.

Electrode gap check – use a wire type gauge for best results.

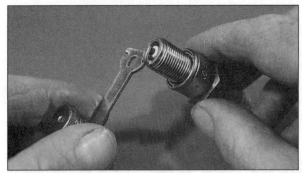

Electrode gap adjustment – bend the side electrode using the correct tool.

Normal condition – A brown, tan or grey firing end indicates that the engine is in good condition and that the plug type is correct.

Ash deposits – Light brown deposits encrusted on the electrodes and insulator, leading to misfire and hesitation. Caused by excessive amounts of oil in the combustion chamber or poor quality fuel/oil.

Carbon fouling – Dry, black sooty deposits leading to misfire and weak spark. Caused by an over-rich fuel/air mixture, faulty choke operation or blocked air filter.

Oil fouling – Wet oily deposits leading to misfire and weak spark. Caused by oil leakage past piston rings or valve guides (4-stroke engine), or excess lubricant (2-stroke engine).

Overheating – A blistered white insulator and glazed electrodes. Caused by ignition system fault, incorrect fuel, or cooling system fault.

Worn plug – Worn electrodes will cause poor starting in damp or cold conditions and will also waste fuel.

6.6a Locate the auto-advance unit dowel

6.6b The hexagonal washer locates with the pegs on the cam

6   Beware of over-tightening the spark plugs, otherwise there is risk of stripping the threads from the aluminium alloy cylinder heads. The plugs should be sufficiently tight to seat firmly on their copper sealing washers, and no more. Use a spanner which is a good fit to prevent the spanner from slipping and breaking the insulator.

7   If the threads in the cylinder head strip as a result of over-tightening the spark plugs, it is possible to reclaim the head by the use of a Helicoil thread insert. This is a cheap and convenient method of replacing the threads; most motorcycle dealers operate a service of this nature. It must be done well, to maintain thermal conductivity between the plug and the cylinder head.

8   Make sure the plug insulating caps are a good fit and have their rubber seals. They should also be kept clean to prevent tracking. These caps contain the suppressors that eliminate both radio and TV interference. Re-connect the plug leads to the correct plug, each lead is clearly marked. The cylinders are numbered 1 - 4 from left to right.

9   Fault diagnosis: ignition system

Symptom	Cause	Remedy
Engine will not start	Faulty ignition switch	Operate switch several times in case contacts are dirty. If lights and other electrics function, switch may need renewal.
	Short circuit in wiring	Check whether fuse is intact. Eliminate fault, before switching on again.
	Completely discharged battery	If lights do not work, remove battery and recharge.
Engine misfires	Faulty capacitor in ignition circuit	Replace capacitor and re-test.
	Fouled spark plug	Replace plug and have original cleaned.
	Poor spark due to generator failure and discharged battery	Check output from generator. Remove and recharge battery.
Engine lacks power and overheats	Retarded ignition timing	Check timing and also contact breaker gap. Check whether auto-advance mechanism has jammed.
Engine 'fades' when under load	Pre-ignition	Check grade of plugs fitted: use recommended grades only.

Chapter 4 Frame and forks

Contents

Specifications

	CB400F	CB550	CB550F
Frame	Semi-duplex cradle	Duplex cradle	Duplex cradle
Wheelbase	1355 mm (53.3 in.)	1405 mm (55.5 in.)	1405 mm (55.5 in.)
Fork caster angle ...	63° 30'	64°	—
Fork trail	85 mm (3.3 in.)	105 mm (4.1 in.)	107 mm (4.2 in.)
Forks		Telescopic, with two-way hydraulic damping	
Fork travel	114.5 mm (4.5 in.)	121 mm (4.8 in.)	—
Spring free length, min.	450 mm (17.717 in.)	425 mm (16.73 in.)	—
Fork slider inside diameter, max.	33.18 mm (1.3063 in.)	—	—
Fork stanchion outside diameter, min.	32.875 mm (1.2944 in.)	—	—
Rear suspension ...		Swinging arm, controlled by two-way hydraulic dampers	
Spring free length, min.	190 mm (7.480 in.)	205 mm (8.070 in.)	190 mm (7.480 in.)
Swinging arm bush inside diameter, max.	21.7 mm (0.8543 in.)	21.8 mm (0.858 in.)	21.7 mm (0.8543 in.)
Swinging arm pivot outside diameter, min.	21.35 mm (0.8406 in.)	21.4 mm (0.342 in.)	21.35 mm (0.8406 in.)
Rear suspension, travel ...	79.0 mm (3.1 in.)	77.3 mm (3.0 in.)	79.0 mm (3.1 in.)
Front fork oil capacity ...	160 - 165 cc (5.6 - 5.8 oz)	155 - 165 cc (5.3 - 5.6 oz)	165 - 170 cc (5.6 - 5.8 oz)
Refill, after draining ...	145 - 150 cc (4.8 - 4.9 oz)	140 - 145 cc (4.7 - 4.8 oz)	150 - 155 cc (4.9 - 5.0 cc)

Torque wrench settings	kg cm	lb ft
Steering head nut	800 - 1200	57.9 - 86.9
Upper pinch bolt	180 - 230	13.1 - 16.7
Lower pinch bolt	180 - 230	13.1 - 16.7
Handlebar clamp bolt	180 - 230	13.1 - 16.7
Rear fork pivot bolt	550 - 700	39.8 - 50.7

Front wheel spindle nut	450 - 550	32.6 - 39.8
Front wheel spindle clamp nut	180 - 230	13.1 - 16.7
Rear wheel spindle nut	800 - 1000	57.9 - 72.4
Footrest fixing nut	450 - 550	32.6 - 39.8

1 General description

The frame of the three models covered by this manual is similar in having a pressed steel spine. The difference lies in the single down tube frame of the CB 400F model and the duplex cradle frame of the CB 550 models.

The spine consists of two halves, welded longitudinally, and forming a large box gusset at the steering head. The single down tube of the CB 400F frame branches out on either side of the oil filter whereas the duplex down tubes of the CB 550 models frame are braced to the spine, below the steering head. In both cases, the duplex cradle sweeps under the engine and up to the rear suspension mountings. At the rear of the spine, two tubes branch out horizontally to join the duplex cradle tubes, continuing rearwards to support seat and mudguard.

Box gussets between these horizontal seat rails and the duplex cradle, form the rear suspension top mountings. Bracing tubes from the seat rails, near their junction with the spine, run down to join the duplex cradle above the swinging arm pivot. Box gussets between these tubes form the pivot mounting. A bracing tube below the pivot provides a mounting for the engine and centre stand. The upper rear left-hand engine mounting is formed by extensions of the swinging arm pivot gussets; a detachable plate on the right allows engine removal. The engine is fixed at the front by engine plates, and by bolts directly to the duplex cradle.

A pressed steel bridge braces the seat rails at the suspension mounting points and carries the mudguard etc.

The rear swinging arm is tubular, with flattened and slotted ends on the CB 400F model, and welded-in fork ends on the CB 550 models. The pivot tube is box gusseted to the swinging arms. The plain bearing is grease lubricated.

Rear suspension is controlled by hydraulic units with two-way damping. The telescopic front forks also have two-way hydraulic damping.

The fork legs are held in the top and bottom fork yokes by pinch bolts. The aluminium alloy lower fork legs are unbushed, and have the damper tube attached to them, at the bottom. On the CB 550 models only, there is an additional damper rod attached to the fork filler nut. A lip seal at the top, protected by a rubber gaiter, prevents ingress of water and grit.

The steering head employs cup and cone ball bearings, top and bottom.

2 Front forks: removal from frame

1 It is unlikely that the front forks will need to be removed from the frame as a complete unit, unless the steering head bearings require attention or the forks are damaged in an accident. It will be necessary to remove the complete hydraulic assembly as a preliminary. It is a sealed system and can be removed in this fashion, thus obviating the need to drain, refill and bleed to eliminate air bubbles.

2 Place the machine on its centre stand with a support under the sump, and remove the front wheel. See Chapter 5, Section 3. Remove the petrol tank to avoid damage.

3 Remove the front mudguard by unscrewing the four bolts which secure the centre bridge to the lower fork legs (the left-hand pair also secure the caliper pivot block). Next remove the two bolts securing the rear lower stay to the lower fork leg, and on the CB 400F only the two which secure the front stay. On the CB 500 models, these bolts are rubber bushed with flanged sleeves. On the CB 400F model, the two bolts securing the caliper pivot and mudguard bridge have a flat and a spring washer.

4 Unhook the hydraulic hose from the mudguard clip on the left-hand side, and the speedometer cable from the two clips on the right-hand side. Pull the guard clear of the forks.

5 Remove the caliper pivot nut on the CB 400F model or the lower bolt on the CB 550 models, followed by the disc guard (not fitted to the CB 550). Next remove the adjuster nut and screw - replace on the caliper bracket together with the spring. Tie the caliper assembly to a convenient point on the frame. If the pivot pin is taken out of the caliper bracket, note the 'O' ring seals. **Do not** operate the front brake as fluid may escape.

6 Disconnect the battery negative (earth) lead. Pull out the plug to the rear of the ignition switch (not on CB 550 models). Unscrew the knurled nuts and withdraw both speedometer and rev-counter cables. Take out the two or three cross-head screws with washers and top hat bushes, and pull off the headlamp unit. Disconnect the bulb holders and put the headlamp aside. Disconnect all wires inside the headlamp shell. Replace the screw etc. in their tapped holes to avoid loss.

7 On the CB 400F model, the instrument bracket is fixed to the upper fork yoke by the two long nuts which also secure the headlamp frame. On the CB 550 models, two ordinary hexagonal nuts secure the instrument bracket only. Remove the bracket complete with instruments and warning lamp console. Pull the wires out of the headlamp shell.

On the CB 400F model only, remove the third lower headlamp frame bolt and take off the headlamp and flasher assembly. All three headlamp frame bolts have grommets with flanged sleeves.

8 Pull off the front stop switch wires. Unscrew the single bolt fixing the hydraulic hose junction to the lower fork yoke. On CB 400F model this also secures the lower headlamp fixing bracket.

9 Remove the right-hand mirror and unscrew the master cylinder clamp bolts. Remove the master cylinder and hydraulic assembly complete. If this is kept upright, and the lever is not depressed, the system will not require bleeding.

10 CB 400F and CB 550F models only: Unscrew the two hexagonal bolts fixing the ignition switch to the underside of the upper fork yoke and remove the switch.

11 Unscrew the handlebar clamp bolts, and remove the handlebar with switches; it may be left attached to the cables, and laid on

2.3 Left-hand mudguard bridge bolts also secure caliper pivot

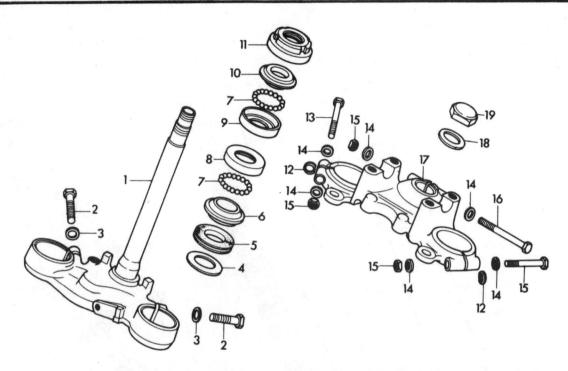

Fig. 4.1. Steering column, CB 550F model (CB 400F similar)

1	Lower fork yoke	7	Ball bearings - 37 off	13	Upper pinch bolt - 2 off	16	Steering column pinch bolt
2	Lower pinch bolt - 2 off	8	Lower bearing cup	14	Flat washer - 4 off (CB 400F)		(CB 550 models only)
3	Flat washer - 2 off	9	Upper bearing cup		- 6 off (CB 550)	17	Upper fork yoke
4	Dust seal washer	10	Upper bearing cone	15	Nut - 2 off (CB 400F) -	18	Flat washer
5	Dust seal	11	Adjuster nut		3 off (CB 550)	19	Steering column top nut
6	Lower bearing cone	12	Spacer washer - 2 off				
			(CB 550 models only)				

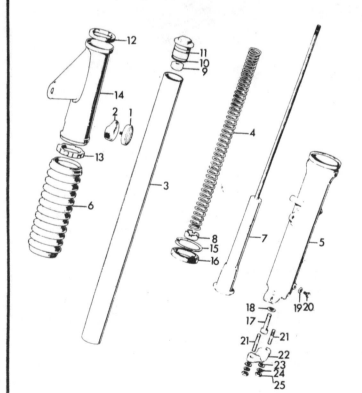

Fig. 4.2. Front forks, CB 550 model

1 Reflector
2 Reflector base
3 Fork stanchion
4 Fork spring
5 Fork slider
6 Gaiter
7 Damper assembly
8 Spring seat
9 Damper rod locknut
10 O-ring seal
11 Fork filler cap
12 Fork shroud seat-top
13 Fork shroud seat-bottom
14 Fork shroud
15 Circlip
16 Oil seal
17 Hexagon socket cap
screw
18 Copper washer
19 Copper washer
20 Filler plug
21 Stud - 2 off
22 Spindle clamp
23 Flat washer - 2 off
24 Spring washer - 2 off
25 Nut - 2 off

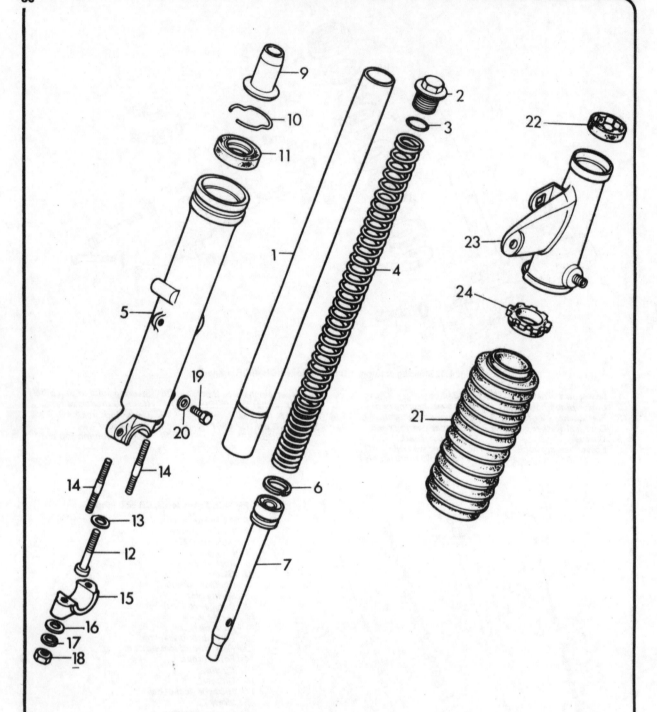

Fig. 4.3. Front forks, CB 400F and CB 550F models

1	Fork stanchion	7	Damper rod	13	Copper washer	20	Copper washer
2	Fork stanchion filler cap	8	Rebound spring (CB 400F model only)	14	Stud - 2 off	21	Fork gaiter (CB 550F shown)
3	O-ring seal	9	Oil lock	15	Spindle clamp	22	Fork cover, shock absorber, top) CB) 550F
4	Fork spring	10	Circlip	16	Flat washer - 2 off	23	Fork cover) only
5	Fork slider	11	Oil seal	17	Spring washer - 2 off	24	Fork cover, shock absorber, bottom)
6	Damper piston ring	12	Hexagon socket screw	18	Nut - 2 off		
				19	Filler plug		

the frame. Take the wiring harness out of the clip beneath the lower fork yoke.

12 Slacken the pinch bolts for the top yoke. Remove the steering column nut and washer and slacken the steering column pinch bolt (CB 550 models only). Take off the top yoke; and the headlamp complete with brackets and flashers on the CB 550 models only. Note the position of the rubber mounting bushes.

13 Unscrew the steering column adjuster nut, but hold the fork in position to avoid losing the balls out of the bottom bearing. Drop the forks out of the steering head, using a rag to catch the lower bearing balls. With the care, the upper bearing balls will remain in position until the upper cone is lifted out, when they drop down the steering column - so be ready to catch them.

14 Clean the grease from the steering head cups, and from the steering column. It is not necessary to remove the cups for checking, but for renewal the old ones must be driven out carefully with a wooden drift. Clean the housing carefully.

3 Front forks: dismantling

1   As front fork action occurs the progressively wound spring inside each fork leg is compressed between the fillter cap and, the top of the damper tube. The damping arrangement differs slightly between that of the CB 550 model and the CB 400F and the CB 550F models, which share a common fork assembly. The damper passes through a bush in the bottom of each fork leg. On the CB 400F and CB 550F models the piston is attached to the damper and slides inside the fork leg. On the CB 550 model a rod attached to the filler cap carries a piston that slides within the damper.

Damping action is similar. As the fork extends, the oil contained in each fork leg is compressed between the damper piston and the bush in the end of the fork leg (CB 400F and CB 550F models) or in the end of the damper (CB 550 model). The oil can escape only through small metering holes in the damper.

On rebound, a similar action takes place and oil flows back into the reservoir around the damper.

2   If only the fork leg require attention, it is not necessary to remove the complete fork assembly as described in the preceding section, but only the front wheel, mudguard and brake caliper unit on the left-hand lower fork leg. It is not necessary to dismantle the hydraulic system, simply tie the caliper to a convenient point on the frame. Since there are two types of damper fitted, these will require different dismantling procedures; but removal of the fork leg as a complete unit requires an identical technique. Deal with each fork leg independently so that parts cannot be interchanged.

3   Note the position of the fork legs in the top yoke - the edge of the chamfer on each tube coincides with the top face of the yoke. Loosen. but do not remove each fork leg filler cap. It is easier to do this now, while the fork leg is clamped firmly. Slacken the top pinch bolt and the lower pinch bolt. Pull each fork leg out of the yokes. Inevitably, they will have seized in place, through rust. Expand the clamp **carefully** with a wedge and use penetrating oil, if still stubborn.

4   Working on one fork leg at a time, remove the fork drain screw and pump the fork to eject all oil. Pull off the fork gaiter.

5   Hold the lower fork leg very carefully in a vice with soft jaws; do not overtighten the vice and distort the fork leg. Unscrew the hexagon socket cap screw in the bottom of the fork leg, **then** withdraw the stanchion filler cap. This prevents the damper from turning.

6   Remove the filler cap complete with its 'O' ring seal. On the CB 550 model, where the damper rod is attached to the filler cap, it is necessary to lift the cap and fit a spanner onto the damper rod locknut on the underside, so that the cap may be unscrewed from the damper rod. Withdraw the fork spring, and the lower spring seat (CB 550 model only). The spring is progressively wound, that is the coils are closer together at one end. This is at the top end on the CB 550 model and the bottom end on the CB 400F and CB 550F models. The difference is

3.3a Chamfer on fork leg aligns with face of top fork yoke

3.3b Slacken pinch bolt on lower yoke

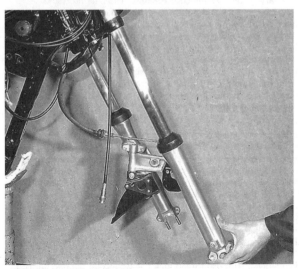

3.3c Withdraw the fork leg

slight in the case of the latter models and notice should be taken of the correct orientation when the spring is withdrawn.

7   Pull the stanchion and damper assembly from lower fork leg. CB 400F and CB 550F models: Push the damper out through the top end of the stanchion. Note that there is an additional spring under the damper piston on the CB 400F model only. CB 550 model: Pull the damper and damper rod out from the bottom end of the stanchion.

8   Remove the internal circlip, or wire circlip which retains the oil seal in the lower fork leg and prize out the oil seal.

4   Front forks: examination and renovation

1   The item most likely to wear, or to be damaged by dirt, is the unbushed lower fork leg. Check the bore for wear or scoring. The lower fork leg must be renewed if any is present. Also check the sliding surface of the stanchion in the same way.

2   Check the damper piston ring for wear on the CB 400F and CB 550F models, or the fit of the damper piston in the bore (CB 550 model).

3   Measure the length of the suspension spring, when unloaded. If compressed, the springs in each fork leg should be renewed as a matched pair.

4   Check the oil seal for scratches or damage. It is advisable to renew the seal in any event. The rudimentary gaiter on the CB 400F model should be a tight fit round the stanchion. If water or grit gain access to the oil seal, it will wear rapidly. For the same reason, there must be no cracks or holes in the convoluted fork gaiter on the other models. Check the condition of the 'O' ring seal on the fork leg filler cap.

5   If the forks have been hit in an accident, check that there are no cracks in the alloy lower fork legs. Check the straightness of each stanchion by rolling it along a flat surface. Do not try to straighten bent stanchions, only renew them. There is no means of knowing if they have been overstressed.

5   Steering head: examination and renovation

1   The only likely damage to the steering head bearings is through maladjustment. They are unlikely to wear out. If, before dismantling the steering head, the forks appear to 'index' in one position when turned, the cups and cones are probably indented and need to be renewed.

2   When dismantled, examine the cups and cones (it is not necessary to remove the cups for this). The bearing tracks should be smooth and polished. If cracked or indented, renew them. Also, examine the bearing balls, and if marked or discoloured, renew them as a complete set. The steering head dust seal below the lower bearing cup should be examined too, and if necessary, renewed.

3   If the forks have been hit in an accident, examine the upper and lower yokes for distortion or cracks. The most likely place for misalignment is the right angle between the steering column and the lower yoke.

6   Front forks: reassembly and replacing in frame

1   Reassemble in reverse order to dismantling. Absolute cleanliness is essential. On the CB 400F and CB 550F models, compress the damper piston ring carefully before inserting the piston into the stanchion. Use the fork spring to push the damper to the bottom, and replace the fork leg filler cap to prevent the damper from turning when tightening its fixing screw. On all models, use a thread lock compound on the damper fixing socket cap screw. The copper washer on this screw should be renewed. Also use thread lock compound on the damper rod thread of the CB 550 model. Tighten the stanchion filler cap onto the damper rod, then tighten the damper rod locknut. Finally screw the filler cap into the fork leg.

2   Smear the oil seal with fork oil and slide it down over the fork stanchion closed side uppermost. Tap the seal home carefully then fit the circlip. Replace the gaiter.

3   Refit the fork drain plug, with a new copper washer and tighten fully. Refill each fork leg with the correct quantity of fork oil and work the forks several times to remove air locks.

4   Push the stanchion through the lower fork yoke, into the upper yoke. Line up both fork legs to the same level; the bottom edge of the chamfer on each fork leg should coincide with top face of the upper fork yoke. Wipe any oil off of the fork leg. With both fork pinch bolts tightened, the fork leg filler caps may be tightened fully.

5   If the complete fork assembly has been removed, it will be easier to replace the steering head and fork legs separately. Stick the bearing balls into the upper and lower bearing cups with grease. Note that on the CB 550 models, there are nineteen balls in the lower race and eighteen in the upper one. On the CB 400F model, there are eighteen balls in the lower race and nineteen in the upper. Although there will be space left for one more, the space is necessary to prevent them from skidding against one another. Insert the steering column into the head, taking care not to displace any balls. Replace the steering column adjuster nut, and adjust the bearings correctly so that there is no play. Now replace the top fork yoke and steering column nut. Carry on with reassembly of the fork legs as already described.

6   After replacing all parts, and tightening all nuts and bolts fully, check fork action by bouncing the front wheel. Check for oil leaks around the oil seals. Check the steering head adjustment again, if the yokes have been removed.

7   Steering head adjustment

1   The steering head bearings should be adjusted so that the forks turn freely from side to side, with no resistance. If, with the bike on the centre stand and the front wheel in line with the frame, the handlebar is pushed, the forks should fall easily to one side. At the same time, there should be no free play in the bearings. Maladjusted steering head bearings are a frequent cause of handling problems. If the forks seem to 'index' in one position when turned, check the bearing tracks for pitting. If the bearings are too tight, the machine will have a characteristic 'roll' at low speeds.

2   With the machine on its centre stand, grasp the forks on each side, at the bottom. Pull and push on the forks horizontally. No movement should be discernable. If another person is able to help, any movement can be felt by putting ones fingers between the top yoke and the steering head.

3   If the bearings require adjustment, slacken the lower fork yoke pinch bolts, the steering column top nut; and the top yoke pinch bolt found only on the CB 550 models. Using a C-spanner, (not supplied in the tool kit) tighten the adjuster nut beneath the top fork yoke until the bearings are free from play. Tighten all the other bolts and check that the forks still turn freely.

8   Draining and refilling the front forks

1   Remove the fork filler cap and drain plug, and allow the oil to drain into a container. Pump the forks up and down to expel the last drops.

2   Clean and replace the drain plug, using a new sealing washer.

3   Refill the forks with the correct quantity of oil. Do not use any other oil. It is important that each fork leg should contain the same quantity of oil. Work the fork several times to expel air.

4   Clean and replace the filler caps, after checking the 'O' ring seal.

9   Frame: examination and renovation

1   If the machine is stripped for an overhaul, this affords an excellent opportunity to inspect the frame for signs of cracks or

3.5 Unscrew the hexagon socket screw

3.6 Unscrew the fork leg filler cap

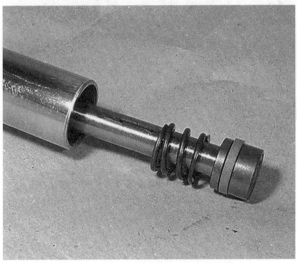

3.7a Push out the damper (CB 400F and CB 550F)

3.7b Damper unit (spring is fitted to CB 400F model only)
(CB 400F and CB550F)

3.8 Oil seal and circlip

8.1 The front fork drain screw

other damage which may have occurred during service. Frame repairs are best entrusted to a frame repair specialist who will have all the necessary jigs and mandrels necessary to ensure correct alignment. This type of approach is recommended for minor repairs. If the machine has been damaged badly as the result of an accident and the frame is well out of alignment, it is advisable to renew the frame without question or, if the amount of money available is limited, to obtain a sound replacement from a breaker's yard.

2   If the front forks have been removed from the machine, it is comparatively simple to make a quick visual check of alignment by inserting a long tube that is a good push fit in the steering head races. Viewed from the front, the tube should line up exactly with the centre line of the frame. Any deviation from the true vertical position will immediately be obvious; the steering head is a particularly good guide to the correctness of alignment when front end damage has occurred. More accurate checking must be carried out when the frame is stripped.

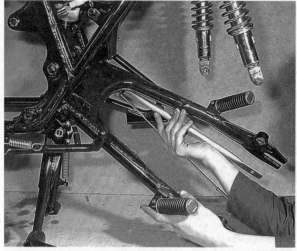

10.4b Take away the swinging arm

10 Swinging arm: removal, examination and renovation

1   Worn swinging arm bearings will cause handling problems, making the rear end twitch and hop, especially when accelerating or shutting off whilst banked over. Play in the bearings may be detected by pulling and pushing horizontally on the rear fork ends. Any play will be magnified by the leverage obtained.

2   To remove the swinging arm fork, firstly remove the rear wheel. See Chapter 5, Section 6.

3   Remove the suspension unit lower mounting bolts. Undo the two chainguard fixing bolts and take off the guard.

4   Unscrew the swinging arm pivot nut and withdraw the long bolt, whilst supporting the swinging arm. Remove the swinging arm with brake torque arm attached.

5   If necessary, remove the torque arm split cotter pin, nut, washer, spring washer and shouldered bolt.

6   The swinging arm bearing end caps have integral seals, the caps pull off. Clean and examine the seals.

7   The long centre pivot pushes out of the swinging arm pivot tube after tapping gently to start it. Clean the bushes in the pivot tube, and the centre pivot and measure their diameters to check wear.

8   Check that the rubber bonded bushes for the suspension units are in good condition. Examine the swinging arm for cracks or damage or rust damage around the gusset welds.

9   Reverse the dismantling procedure when reassembling. Note

10.6a The pivot end cap pulls off ...

10.4a Remove the swinging arm pivot nut

10.6b ... it has an integral seal

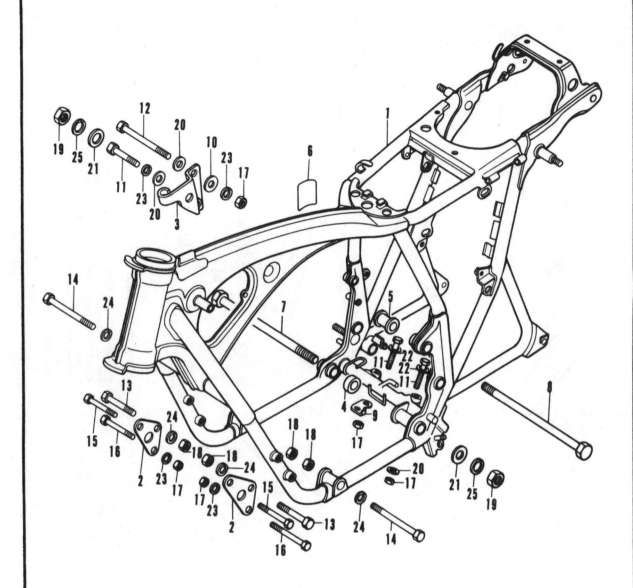

Fig. 4.4. Frame

1 Frame
2 Front engine plate - 2 off
3 Rear engine plate
4 Right engine lower spacer -
   CB 550 models only
5 Left engine upper spacer
6 Battery caution label
7 Rear engine bolt - lower
8 Rear engine bolt - upper
9 Battery breather guide plate -
   CB 550 models only
10 Washer - CB 550 models
   only

11 Bolt - 2 off CB 400F
   3 off CB 550
12 Bolt - CB 550 models
   only
13 Bolt - 1 off CB 400F -
   2 off CB 550 models
14 Bolt - 2 off
15 Bolt - 2 off
16 Bolt - 2 off CB 550
   models only
17 Nut - 2 off - CB 400F -
   7 off CB 550 models
18 Nut - 3 off CB 400F - 4 off
   CB 550 models

19 Nut - 3 off
20 Plain washer - 3 off -
   550 models only
21 Plain washer - 2 off
22 Plain washer - 2 off - CB
   550 models only
23 Spring washer - 2 off CB 400F
   - 6 off CB 550 models
24 Spring washer - 3 off
   CB 400F - 4 off CB 550
   models
25 Spring washer - 2 off

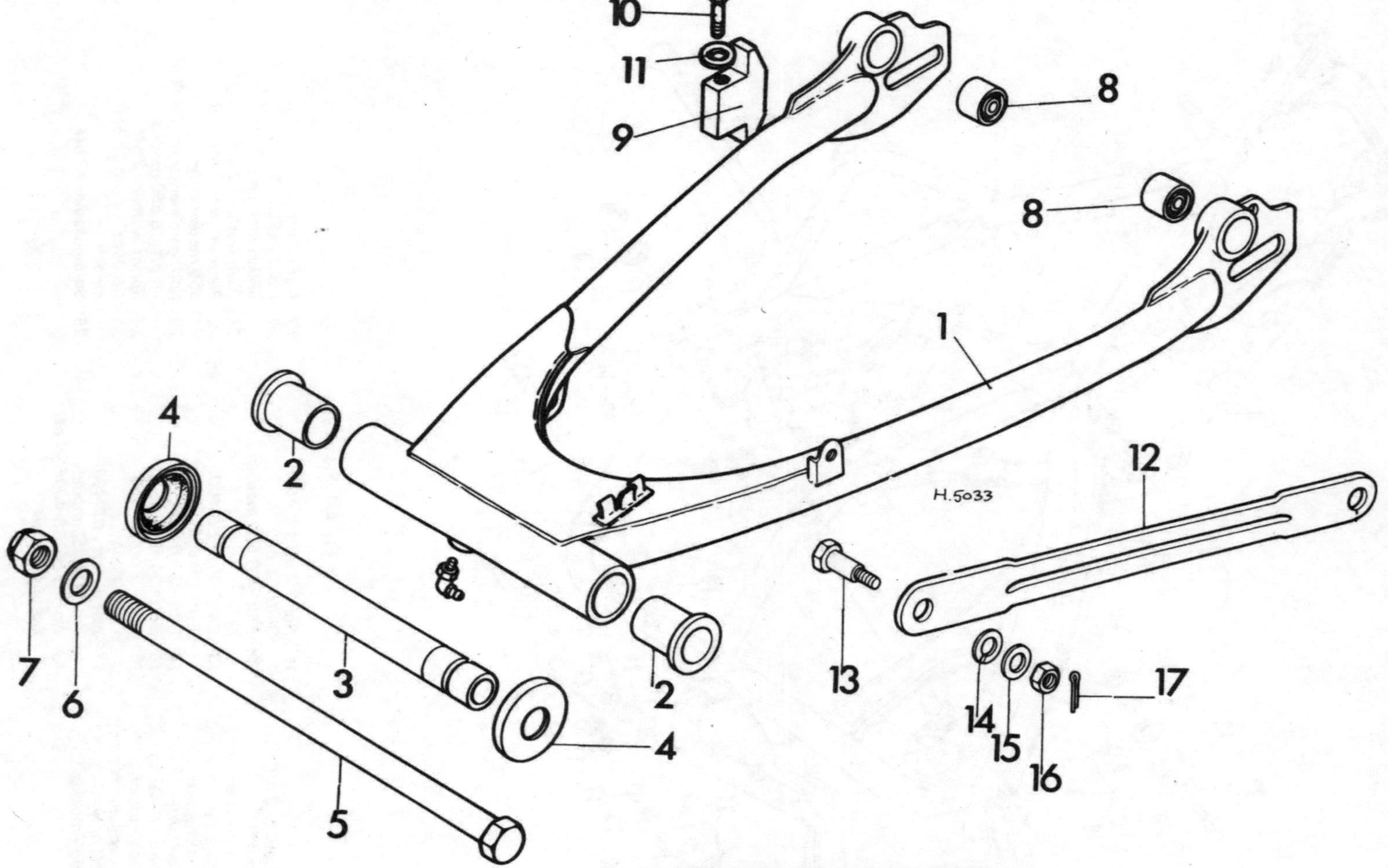

H.5o33

Fig. 4.5. Swinging arm (similar for all models)

1 Rear swinging arm
2 Bush - 2 off
3 Centre pivot
4 Dust seal

5 Pivot bolt
6 Flat washer
7 Self locking nut
8 Rubber bonded bush - 2 off

9 Rear fork closer (CB 400F and CB 550 models) - 2 off
10 Bolt (CB 550 models only) - 2 off
11 Washer (CB 550 model) - 2 off
12 Rear brake torque arm

13 Shouldered bolt
14 Spring washer
15 Flat washer
16 Castellated nut
17 Split pin

that the front end of the chainguard fits into a clip on the pivot gusset. Grease the pivot via the grease nipple above the pivot tube on the CB 400F model, below on the CB 550F model and at the end on the CB 550 model. Check the chain tension.

10 Note: CB 550 models only. If the brake pedal and shaft have been removed for any reason, make sure that the lever to which the front end of the brake rod is attached is **in front of** the swinging arm pivot tube. It is impossible to reposition it after reassembly without removing either the swinging arm again, or the brake return spring, this latter item being difficult to retension in situ.

11 Rear suspension units: examination and renovation

1 The units are adjustable for load in five positions. Turn the adjuster clockwise with the toolkit C-spanner to increase the spring preload. Both units **must** be at the same setting.

2 Examine the damper units for oil leaks. Bounce the rear of the machine. If it does not come to rest after a couple of oscillations, the dampers may be faulty.

3 The dampers are sealed units, and cannot be repaired. If oil leaks are apparent, the oil seal is damaged and the damper requires renewal. It is best to renew **both** damper units, as a matched pair.

4 To remove the damper unit, it is necessary first to take off the lifting handle. This is fixed to the suspension unit top mounting, and to the rearwards frame extensions. Unscrew the suspension unit lower mounting bolt, and pull off the unit.

5 Hold the lower mounting in a vice and depress the spring cover until the split cotters can be removed. It is best to employ a second person to remove the collets whilst the spring is compressed. Remove the cover and spring. It should be possible to feel an equal resistance throughout the travel of the damper rod, in both directions. Check the spring free length.

6 Reverse this procedure to reassemble the suspension unit

12 Centre stand: examination

1 The centre stand pivots on a hollow tube, clamped between two split lugs below the rear engine mounting. The pinch bolts should be checked occasionally, for security and the pivot oiled.

2 Also check that the return spring and linkage is unworn and retracts the stand smartly. If the stand falls whilst the bike is in motion, an accident may result.

13 Prop stand: examination

1 The prop stand is pivoted on a lug under the front left-hand engine mount. Check the security of the pivot bolt and oil occasionally.

2 Check the action of the stand, and make sure that the return spring is not worn or weakened. An accident will almost inevitably result from the prop stand dropping whilst the machine is in motion.

3 To prevent the machine from being ridden away with the prop stand down, American 'F' models incorporate a novel self-retracting device. Check the rubber 'trip' of this device for wear or damage. No part should be worn below the moulded line on the rubber.

4 Check the operation of the stand as follows: Put the machine on the centre stand, and put the prop stand down. Using a spring balance attached to the extreme end of the centre stand, measure the force required to retract the stand. If this force exceeds 2 - 3 kg (4.4 - 6.6 lb), check that the stand pivot bolt is not overtight, or in need of lubrication.

5 To renew the rubber 'trip', take off the bolt. Make sure the sleeve is installed in the fixing hole of the new trip. Fit the trip with the arrow facing outwards. The block should be marked 'over 260 lbs only'.

14 Footrests: examination

1 The front footrests are fitted to the rear engine mounting stud. They are non-adjustable, folding and spring loaded in the horizontal position. The footrest rubber is held to the support with two bolts passing through a plate underneath the rubber.

2 Owing to their construction, it is unlikely that the footrests can be repaired if damaged.

3 To renew a return spring, remove the split cotter pin and pull out the clevis pin from the footrest pivot. The right-hand footrest has a semi-circular leaf spring which also holds the rest vertical for starting. The closed end of this spring locates over a dowel pin on the footrest hanger, and the open end over a second dowel on the footrest.

4 The pillion footrests are folding and non-adjustable. On the CB 400F model they are attached directly to the swinging arm. Check the security of the bolts and pivots.

15 Rear brake pedal: examination

1 The rear brake pedal is attached to a splined pivot passing through a tube welded to the frame below the swinging arm pivot.

2 The height of the pedal may be adjusted using the bolt provided.

3 The pedal is fixed to the splined pivot with a pinch bolt. If the pedal is to be removed, mark the relative positions of pedal and shaft, if not already done. Check the condition of the return spring.

4 CB 550 models only. If the brake pedal pivot and arm are removed, note that the arm to which is attached the front end of the brake rod must be positioned **in front of** the swinging arm pivot tube.

16 Gear lever: CB 400F models only

1 The CB 400F model has an external gearchange linkage to allow better ergonomic positioning of the pedal. The pedal itself is pivoted on a pin welded to the left-hand swinging arm gusset. This pedal operates the gearchange shaft lever by a link with a rod end joint at each end.

2 The gearchange pedal can be taken off, after removing the external circlip and washer from the pivot pin. Also remove the gearchange shaft lever pinch bolt. After noting the position of the gearchange shaft lever on its splines (there are centre punch marks on each item), pull off the complete linkage.

3 The rod end joints are pressed into their levers, and cannot be renewed individually. They should not give trouble if lubricated occasionally. The height of the pedal can be adjusted by slackening the locknuts on the linkage rod, and turning the rod.

17 Dual seat: removing

1 The seat is hinged on the left-hand side of the rear subframe. The hinges consist of clevis pins, secured with split pins. After removing the split pins and the clevis pins, the seat will lift off.

18 Dual seat lock: examination

1 The dual seat lock is actuated by the ignition key. If the lock has to be renewed, try to obtain one with the same key number. The lock is secured to the underside of a bracket on the right-hand subframe by two cross-head screws.

2 The latch and drum of the lock should be oiled occasionally, never oil the internal lock mechanism.

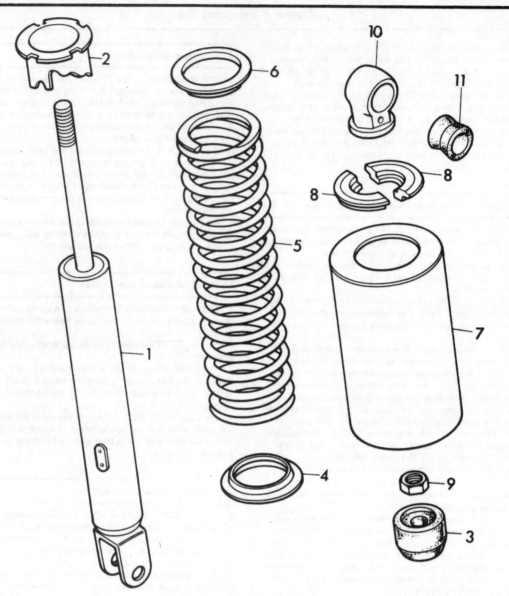

Fig. 4.6. Rear suspension unit

1 Damper unit - 2 off
2 Spring adjuster cam -
   2 off
3 Rubber stop - 2 off

4 Lower spring seating -
   2 off
5 Spring - 2 off
6 Upper spring seating - 2 off

7 Upper shroud - 2 off
8 Spring retaining collets -
   4 off
9 Damper unit nut - 2 off

10 Upper fixing point -
    2 off
11 Rubber bush - 2
    off

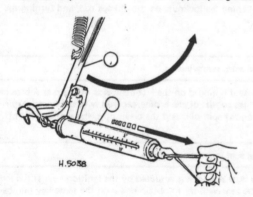

H.5038

Fig. 4.7. Testing prop stand action (US models - CB 400F and
CB 550 F)

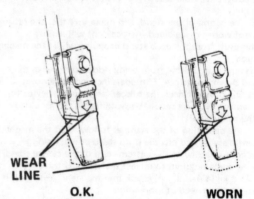

WEAR
LINE

O.K.   WORN

Fig. 4.8. Prop stand retracting trip wear guide (US models -
CB 400F and CB 550F)

19 Steering head lock: examination

1 The steering head lock has been mounted in various positions on the steering head.

On the CB 550 models it is attached to the underside of the lower fork yoke, by a screw and washer. The USA CB 400F and CB 550F models have a lock integral with the ignition switch. On UK CB 400F model, the lock is on the left-hand side of the steering column, retained when in the off position by a rivet and latch plate.

2 If a lock is faulty it must be renewed. Oil **only** the body of the lock occasionally; do not oil the internal mechanism. On the UK CB 400F model, the key will be broken off it it is left in the lock and the steering turned. This will render the lock useless.

20 Instrument drive cables: examination and replacing

1 Drive cables should be examined and lubricated occasionally. The outer sheath should be examined for cracks or damage, the inner cable for broken or frayed strands. Jerky or sluggish instrument movement is generally caused by a faulty cable.

2 Detach the cable at the drive end, and withdraw the inner cable. Clean and examine the cable. Relubricate it with high melting point grease, but do not grease the top six inches of cable, at the instrument end, or grease will work its way into the instrument head and ruin the movement.

3 Re-route the cables as they were originally. Make sure that the steering turns freely.

21 Instrument heads - removing

1 Unscrew the knurled cable nut and disconnect the drive cable. Unscrew the two acorn nuts and remove the large washer, grommet and sleeve. Lift the head and pull out the bulb holders. Remove the head.

2 It is not possible to repair a faulty instrument head. If the instrument fails completely, or moves jerkily, first check the drive cable. If the mileage recorder of the speedometer ceases to function but the speedometer continues to work or vice versa, the instrument head is faulty.

3 Remember that a working speedometer, accurate at 30 mph, is a statutory requirement in the UK.

22 Sidecar - fitting

1 The models covered by this manual have no provision for sidecar attachment, which is not recommended by the manufacturer. However, there is always someone who achieves the impossible and overcomes the fitting problems, hence the inclusion of this addition advice.

2 Good-handling characteristics of the outfit will depend on the accuracy with which the sidecar is aligned. Provided the toe-in and lean-out are within prescribed limits, good handling characteristics should result, leaving scope for other minor adjustments about which opinions vary quite widely.

3 To set the toe-in, check that the front and rear wheels of the motorcycle are correctly in line and adjust the sidecar fittings so that the sidecar wheel is approximately parallel to a line drawn between the front and rear wheels of the machine. Re-adjust the fittings so that the sidecar wheel has a slight toe-in toward the front wheel of the motorcycle, as shown in Fig. 4.9a. When the amount of toe-in is correct, the distance 'B' should be from 3/8 inch to 3/4 inch less than the distance at 'A'.

4 Lean-out is checked by attaching a plumb line to the handle bars and measuring the distance between 'C' and 'D' as shown in Fig. 4.9b. Lean-out is correct when the distance 'C' is approximately 1 inch greater than at 'D'.

5 Note that if a sidecar is fitted to any of the four cylinder models covered by this manual, it will invalidate the Honda guarantee.

23 Cleaning the machine

1 If possible, the machine should be wiped down immediately after use in the wet, so that it is not garaged in a rust-promoting condition.

2 Before regular cleaning, wash off dirt with plenty of water, and allow the machine to dry. Use a wax polish on painted parts, and a proprietary chrome cleaner such as Solvol Autosol, on plated and aluminium alloy items. Note that some alloy parts on these motorcycles have a lacquered finish which needs wiping only.

3 If any part of the machine is caked with an oily film, use a cleaner such as Gunk, in accordance with the instructions. Keep water out of the carburettor and ignition components. After washing down the machine, re-oil exposed control cables, bearings and the chain.

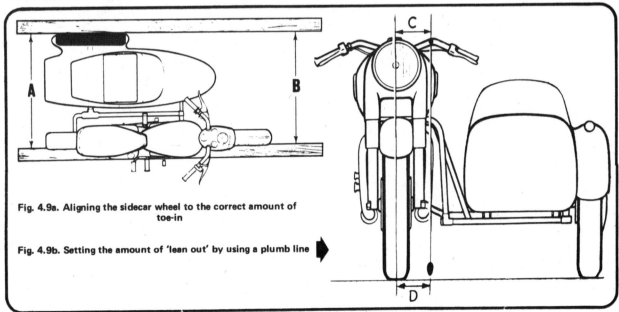

Fig. 4.9a. Aligning the sidecar wheel to the correct amount of toe-in

Fig. 4.9b. Setting the amount of 'lean out' by using a plumb line

24 Fault diagnosis: frame and forks

Symptom	Cause	Remedy
Machine veers to left or right with hands off handlebars	Incorrect wheel alignment Bent forks Twisted frame	Check and realign. Check and renew. Check and renew.
Machine rolls at low speeds	Overtight steering head bearings	Slacken and re-test.
Machine judders when front brake is applied	Slack steering head bearings	Tighten until all play is taken up.
Machine pitches badly on uneven surfaces	Ineffective fork dampers Ineffective rear suspension units	Check oil content. Check damping action.
Fork action stiff	Fork legs out of alignment (twisted yokes)	Slacken yoke clamps, front wheel spindle and fork top bolts. Pump forks several times, then tighten from bottom upward.
Machine wanders. Steering imprecise, rear wheel tends to hop	Worn swinging arm pivot	Dismantle and renew bushes and pivot shaft.

Chapter 5 Wheels, brakes and tyres

Contents

Specifications

Tyres	CB400F	CB550	CB550F
Front	3.00 S18 (4PR)	3.25 S19 (4PR)	3.25 S19 (4PR)
Rear	3.50 S18 (4PR)	3.75 S18 (4PR)	3.75 S18 (4PR)
Tyre pressures solo			
Front	26 psi (1.8 kg/cm^2)	26 psi (1.8 kg/cm^2)	25 psi (1.7 kg/cm^2)
Rear	28 psi (2.0 kg/cm^2)	28 psi (2.0 kg/cm^2)	28 psi (2.0 kg/cm^2)
Tyre pressures two-up			
Front	26 psi (1.8 kg/cm^2)	26 psi (1.8 kg/cm^2)	28 psi (2.0 kg/cm^2)
Rear	36 psi (2.5 kg/cm^2)	36 psi (2.5 kg/cm^2)	36 psi (2.5 kg/cm^2)
Front brake		Hydraulically operated single disc	
Lining area	38 cm^2 (5.9 in.^{2})	—	—
Brake disc thickness			
min.	6 mm (0.236 in.)	6 mm (0.236 in.)	6 mm (0.236 in.)
Brake disc face runout			
max.	0.3 mm (0.012 in.)	0.3 mm (0.012 in.)	0.3 mm (0.012 in.)
Caliper cylinder inside diameter			
max.	38.215 mm (1.5045 in.)	38.215 mm (1.5045 in.)	38.215 mm (1.5045 in.)
Caliper piston outside diameter			
min.	38.105 mm (1.5002 in.)	38.105 mm (1.5002 in.)	38.105 mm (1.5002 in.)
Master cylinder inside diameter			
max.	14.055 mm (0.533 in.)	14.055 mm (0.533 in.)	14.055 mm (0.533 in.)
Master cylinder piston outside diameter			
min.	13.940 mm (0.488 in.)	13.940 mm (0.488 in.)	13.940 mm (0.488 in.)
Rear brake		Rod operated single leading shoe drum	
Minimum lining thickness	2.5 mm (0.098 in.)	2.0 mm (0.09 in.)	2.0 mm (0.09 in.)
Brake drum diameter			
max.	161 mm (6.3386 in.)	181 mm (7.125 in.)	181 mm (7.125 in.)
Wheel bearing end play			
max.	0.1 mm (0.004 in.)	0.1 mm (0.004 in.)	0.1 mm (0.004 in.)

Wheel bearing radial play			
max.	0.05 mm (0.002 in.)	0.05 mm (0.002 in.)	0.05 mm (0.002 in.)
Wheel rim lateral run-out			
max.	2.0 mm (0.079 in.)	2.0 mm (0.079 in.)	2.0 mm (0.079 in.)

Torque wrench settings

Front wheel spindle nut	450 - 550 kg cm	550 - 650 kg cm	450 - 550 kg cm
... ...	(36.6 - 39.8 lb ft)	(39.8 - 47.0 lb ft)	(36.6 - 39.8 lb ft)
Front spindle clamp nut	180 - 230 kg cm	180 - 230 kg cm	180 - 230 kg cm
... ...	(13.1 - 16.7 lb ft)	(13.0 - 16.6 lb ft)	(13.1 - 16.7 lb ft)
Rear wheel spindle nut	800 - 1000 kg cm	800 - 1000 kg cm	800 - 1000 kg cm
... ...	(57.9 - 72.4 lb ft)	(57.8 - 72.3 lb ft)	(57.9 - 72.4 lb ft)
Brake torque arm nuts	180 - 230 kg cm	180 - 230 kg cm	180 - 230 kg cm
... ...	(13.1 - 16.7 lb ft)	(13.0 - 16.6 lb ft)	(13.1 - 16.7 lb ft)
Front brake caliper clamp bolts	340 - 400 kg cm	340 - 400 kg cm	340 - 400 kg cm
... ...	(24.6 - 28.9 lb ft)	(24.6 - 28.9 lb ft)	(24.6 - 28.9 lb ft)
Rear sprocket nuts	400 - 500 kg cm	300 - 400 kg cm	300 - 400 kg cm
... ...	(29.0 - 32.2 lb ft)	(21.7 - 28.9 lb ft)	(21.7 - 28.9 lb ft)

1 General description

Both front and rear wheels are fitted with steel rims, laced to alloy hubs which are supported on two ball journal bearings.

The rear wheel is not quickly detachable, requiring detachment of the final drive chain prior to removal. The single leading shoe drum brake is rod operated.

The front hydraulic disc brake has a self-adjusting floating caliper, pivoted on the fork slider. The disc is of stainless steel to resist corrosion. The hydraulic master cylinder and reservoir is mounted in an exposed position on the handlebar, where it is operated directly by the lever.

Final drive is by an exposed, unlubricated, roller chain. There is a rubber bushed shock absorber in the rear hub.

2 Front wheel: examination

1 Place the machine on its centre stand, with a block under the sump, so that the front wheel is clear of the ground.

2 Spin the wheel and check for rim alignment. Small irregularities can be corrected by tightening the spokes in the affected area, although a certain amount of experience is necessary if over-correction is to be avoided. Any 'flats' in the wheel rim should be evident at the same time. These are more difficult to remove with any success, and in most cases the wheel will have to be rebuilt on a new rim. Apart from the effect on stability, especially at high speeds, there is much greater risk of damage to the tyre beads and walls if the machine is ridden with a deformed wheel.

3 Check for loose or broken spokes. Tapping the spokes is the best quide to tension. A loose spoke will produce a quite different note and should be tightened by turning the nipple in an anticlockwise direction. Always check for run-out by spinning the wheel again.

If a spoke has been tightened, the tyre should be removed to ensure that the spokes do not protrude beyond the nipple, when it may puncture the tube. File or grind off the protruding ends.

4 Grasp the wheel at its periphery, and push and pull on the rim, to check for play in the wheel bearings.

3 Front wheel: removal and replacement

1 With the front wheel supported clear of the ground, remove the cross-head screw which retains the speedometer cable to its drive gearbox, on the right-hand side of the hub. Pull out the cable, and replace the screw to prevent its loss.

2 Undo the 14 mm nuts which hold each spindle clamp to the lower fork legs, supporting the wheel whilst doing so. Remove the spindle clamps with nuts, flat and spring washers. Allow the wheel to drop and roll it out complete with spindle.

3 **Do not operate** the front brake while the wheel is removed, since fluid can escape from the hydraulic system. To avoid such an accident, it is a good idea to put a piece of wood 7 mm (¼ inch) thick between the brake pads.

4 To replace the wheel, position it in the forks and replace the spindle clamps. The brake side clamp locates in a groove on the spindle nut and should be **tightened first**. Also, the nut at the front of the clamps should be **tightened fully** first, followed by the rear nut (on both sides). Reconnect the speedometer cable and ensure that it is correctly routed.

5 Spin the wheel to make sure that it revolves freely, and check the brake operation. Check that all nuts and bolts are fully tightened.

4 Front wheel bearings: examination and renovation

1 Prevent the spindle from rotating by inserting a tommy bar, and unscrew the sleeve nut on the brake disc side. Remove the nut and the spacer, then extract the spindle. Pull off the speedometer drive gearbox.

2 Long bolts (four on the CB 400F model, six on the CB 550 models) retain both the disc and the right-hand bearing cover. Undo the nuts, after bending down the tab washers on the CB

3.1 Pull out speedometer drive cable

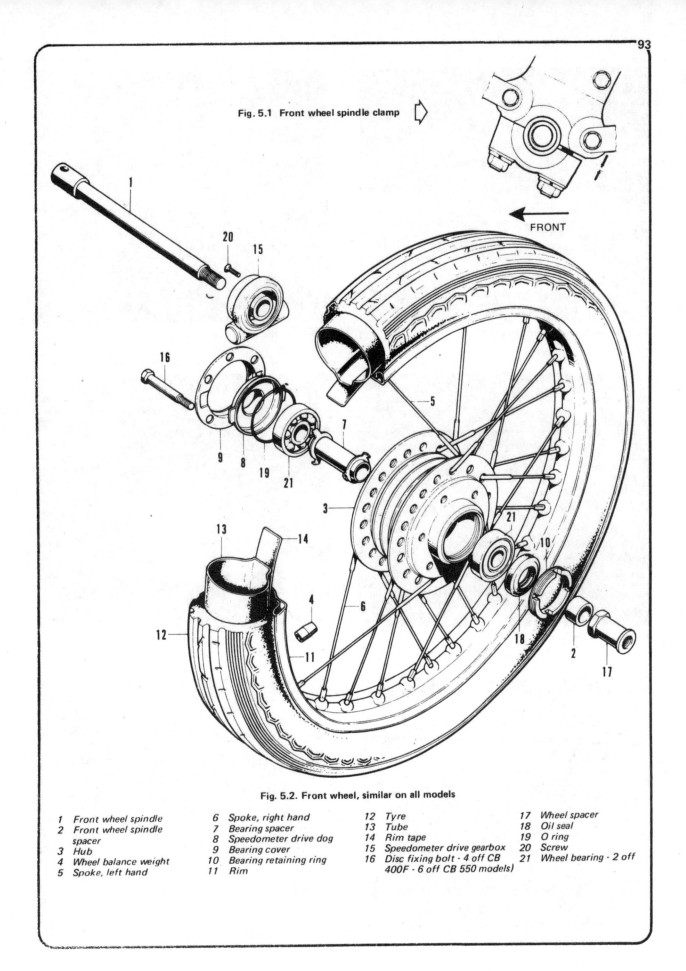

Fig. 5.1 Front wheel spindle clamp

FRONT

Fig. 5.2. Front wheel, similar on all models

1	Front wheel spindle	6	Spoke, right hand	12	Tyre	17	Wheel spacer
2	Front wheel spindle	7	Bearing spacer	13	Tube	18	Oil seal
	spacer	8	Speedometer drive dog	14	Rim tape	19	O ring
3	Hub	9	Bearing cover	15	Speedometer drive gearbox	20	Screw
4	Wheel balance weight	10	Bearing retaining ring	16	Disc fixing bolt - 4 off CB	21	Wheel bearing - 2 off
5	Spoke, left hand	11	Rim		400F - 6 off CB 550 models)		

3.2a Remove the wheel spindle clamp

3.2b Allow the wheel to drop out

4.1 Withdraw the spindle

4.2 Speedometer drive and bearing cover (CB 400F)

4.3a Remove the bearing cover

4.3b Note O-seal on spigot, and bearing seal

550 models and remove the disc. Mark it and the hub so that it eventually is replaced in the same position, to preserve balance.

3   Remove the long bolts and the bearing cover, followed by the speedometer drive dog. Note the 'O' ring seal over the spigot on the right-hand side of the hub.

4   Unscrew the bearing retaining ring on the disc side of the hub. This is of aluminium alloy, staked into the hub, and very easily damaged by punching round. The grease seal also is pressed into this ring and will be damaged too. Use only a properly fitting tubular tool, unless both items are to be renewed.

5   The disc side bearing may be driven out from the opposite side, using a soft metal drift which locates on the inner diameter of the bearing spacer. When this bearing and the spacer, have been removed, the right-hand bearing can be driven out, using the same drift.

6   Remove excess grease from the bearings, spacer and inside the hub. Clean the bearing housings in the hub. Wash the bearings in white spirit to remove all grease. **Do not spin** a dry bearing. If the bearings show more than very slight radial play, or roughness when rotated, they should be renewed.

7   A scorched, glazed appearance of the bearing housing and the outer diameter of the bearing will indicate that it has been rotating in the hub. If damage to the hub is not too great, replace the bearing using Loctite or a similar bearing adhesive to effect a cure.

8   Pack the bearings with grease before refitting. **Do not** over grease, as this will only raise the running temperature of the bearing, and create drag. Tap the bearings into the hub, not omitting the spacer. Tap only the **outer** ring of a bearing, **never** the inner. A tube is the best tool to use, ensuring that the bearings enter the housings squarely. It is an advantage to freeze the bearings before fitting, to decrease the interference fit. The integral seals must face **outwards.**

9   Check the 'O' ring seal on the hub spigot, and replace all parts in reverse order. Check the condition of the multi-lip seal in the retaining ring and renew it if it is scratched or damaged. Use Loctite or any similar adhesive to lock the ring in position, without staking.

10 On the CB 550 models the speedometer gearbox drive dog must locate in the two recesses in the bearing cover.

11 The speedometer gearbox itself should be engaged with the tongues on the drive dog. Use new tab washers, where fitted. The spindle nut goes on the disc side of the hub, with the hexagon adjacent to the spacer, which passes through the grease seal. Make sure that there are no sharp edges on the spacer before pushing it into the seal. The spindle should be greased before replacement.

4.4 Unscrew the bearing retaining ring

6.3 Remove the spring pin on the torque arm bolt

---

5   Rear wheel: examination

1   Place the machine on its centre stand, with the rear wheel clear of the ground.

2   Spin the wheel to check for misalignment as described for the front wheel (Sections 2.2 and 2.3). It is easier to do this with the chain removed from the rear wheel sprocket.

---

6   Rear wheel: removal and replacement

1   The rear wheel is not quickly detachable. The chain must first be taken off and the brake removed with the wheel.

2   Put the machine on the centre stand, with the rear wheel well clear of the ground. Unscrew the rear brake rod adjuster.

3   Remove the spring cotter pin and undo the torque arm nut on the brake back plate. Remove the large washer, rubber washer and the torque arm. Replace the washers nut and pin to avoid loss.

4   Remove the spindle nut split pin and slacken the nut. **Do not** disturb the chain adjusters if a link is fitted to the chain. This will avoid having to readjust the chain later.

6.6 Pull the wheel rearwards (CB 400F)

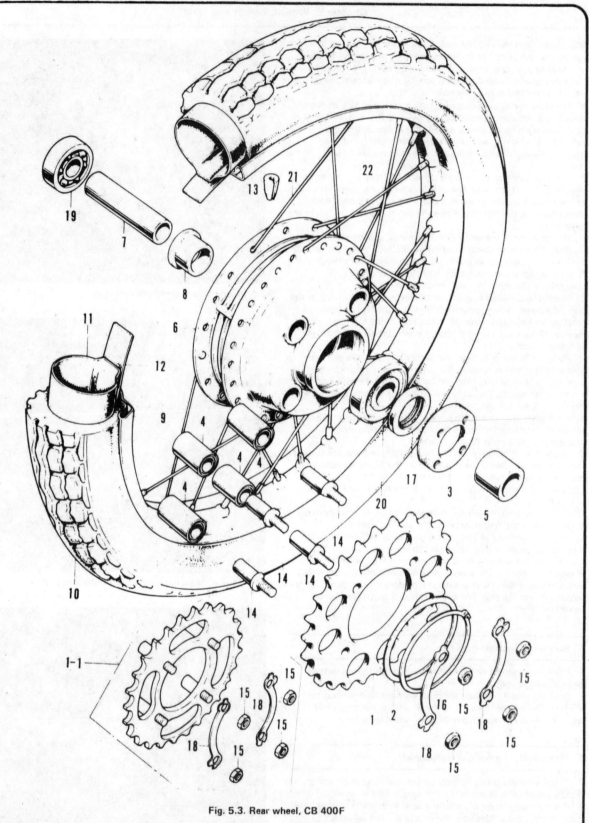

Fig. 5.3. Rear wheel, CB 400F

1	Sprocket	6	Hub	12	Rim tape	17
2	Washer	7	Bearing spacer	13	Balance weight	18
3	Bearing retaining ring	8	Flanged bearing spacer	14	Shock absorber stud -	19
4	Shock absorber bush -	9	Rim		4 off	20
	4 off	10	Tyre	15	Nut · 4 off	21
5	Spindle spacer	11	Tube	16	Circlip	22

1 Sprocket
2 Washer
3 Bearing retaining ring
4 Shock absorber bush - 4 off
5 Spindle spacer
6 Hub
7 Bearing spacer
8 Flanged bearing spacer
9 Rim
10 Tyre
11 Tube
12 Rim tape
13 Balance weight
14 Shock absorber stud - 4 off
15 Nut · 4 off
16 Circlip
17 Grease seal
18 Locking washer - 2 off
19 Ball journal bearing
20 Ball journal bearing
21 Outside spoke - 18 off
22 Inside spoke - 18 off

5 Find the chain connecting link (when fitted), and remove the
spring clip with flat pliers. Take off the side plate, and pull out
the link. Unhook the chain from the rear sprocket. Do not allow
the chain to come off of the gearbox sprocket. Hook it up out
of the dirt. If the chain is of the endless type, the chain adjusters
will have to be slackened off, so that the chain can be slipped off
the sprocket.
6 On the CB 400F model, pull the wheel complete with spindle
and adjusters clear of the fork ends and roll it out between the
fork and mudguard. Remove the trunnion from the brake arm,
and replace it on the brake rod with the spring and adjuster nut.
7 The CB 500 models have closed fork ends. Remove the spindle
nut plus washer completely, and pull out the spindle with the aid
of a tommy bar. Collect the chain adjusters and brake side
spacer. Roll the wheel with brake back plate out between the
fork and the mudguard.
8 Before replacing the rear wheel of the CB 400F model,
assemble the spindle and all spacers, adjusters etc, on the wheel.
9 Position the wheel between the rear fork ends and engage
with the spindle slots. On the CB 550 models, insert the greased
spindle from the drive side, through the adjuster, fork ends,
brake side spacer and finally replace the spindle nut and washer.
Do not tighten.
10 Refit the brake torque arm. It has to locate over the larger
diameter of the torque arm stud. Fit the rubber washer, flat
washer, nut and spring pin.
11 Refit the brake rod, spring and adjuster.
12 Make sure that the wheel is pushed fully forwards, with the
adjusters positioned correctly and tighten the spindle nut.
Replace the split pin, which should be renewed if reused more
than a couple of times.
13 Spin the wheel to ensure freedom of rotation, and check
brake operation. Hook the chain over the sprocket and refit the
spring link. The easiest way to do this is to press the two ends
into adjacent teeth on the sprocket, then insert the link. The
closed end of the spring clip **must** face the direction of travel
of the chain. It may otherwise be forced off.
14 Check the chain tension, brake adjustment and wheel
alignment. Important: Make sure the brake torque arm is
connected and the bolts tightened. If the torque arm comes
loose, a serious accident can result, as the rear brake will lock on
immediately it is applied.

7 Rear wheel bearings: examination and renovation

1 On the CB 400F model remove the spindle nut, washer
chain adjuster and spacer. Pull out the spindle with the drive side
adjuster and spacer.
2 Take off the brake back plate and put aside.
3 Remove the rear sprocket. On the CB 400F model this is
retained by a very large circlip. After removing this circlip, and
the large washer, pull off the sprocket with cush drive studs
attached. The sprocket may be quite tight.
    The CB 550 models sprocket is held by four nuts with tab
washers. Flatten these, remove the nuts and tap off (if necessary)
the sprocket and cover plate, using a piece of wood.
4 Unscrew the bearing retaining ring on the drive side using a
purpose-made tool, for the reasons stated in the front wheel
section (4.4) of this Chapter. The ring may have a right or left-
hand thread depending on the year of manufacture. Some exper-
imentation will be necessary to determine the type of thread
used, and hence the correct direction of rotation for removal.
Knock out the drive side bearing from the brake side, as described
for the front wheel, Section 4.5 of this Chapter.
5 Clean, examine and replace the bearings as for the front
wheel, in Sections 4.6 to 8 of this Chapter.
6 Check the grease seal on the drive side for damage or scratches
and renew if necessary. Tighten the bearing retaining ring fully,
using Loctite or a similar adhesive to lock it in position. On the
CB 400F model the grease seal is fitted before the retaining ring.
On the CB 550 models the seal is pressed into the sprocket

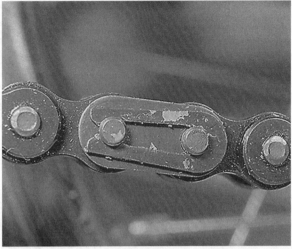

6.13 The closed end **must** face direction of chain travel

7.3 Remove the sprocket circlip (CB 400F)

7.4 Unscrew the bearing retaining ring

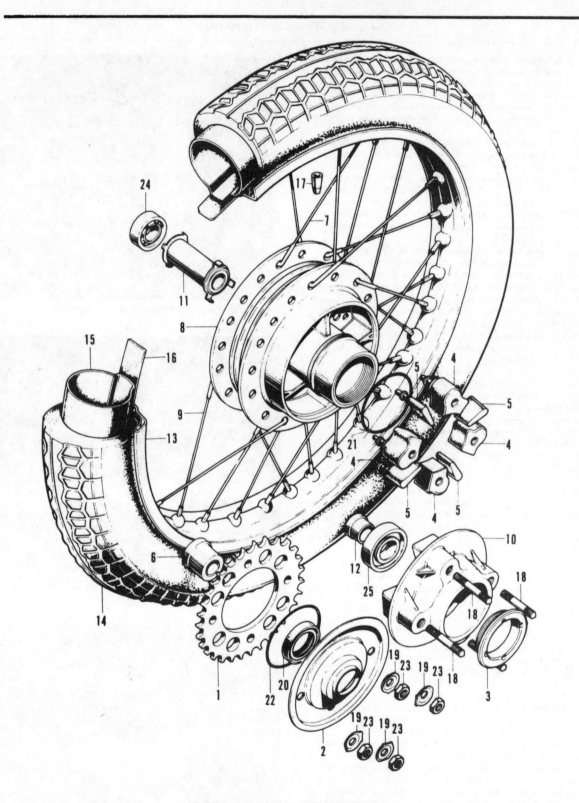

Fig. 5.4. Rear wheel, CB 550 and 550F models

1	Final drive sprocket	6	Spindle spacer	13	Rim	20	Grease seal
2	Sprocket cover	7	Spoke (left-hand) - 20 off	14	Tyre	21	O-ring seal
3	Bearing retaining ring	8	Hub	15	Tube	22	O-ring seal
4	Shock absorber rubber - thick - 4 off	9	Spoke (right-hand) - 20 off	16	Rim tape	23	Nut - 4 off
5	Shock absorber rubber - thin - 4 off	10	Final drive flange	17	Balance weight	24	Ball journal bearing (right-hand)
		11	Bearing spacer	18	Stud - 4 off	25	Ball journal bearing (left-hand)
		12	Inner spacer	19	Tab washer - 4 off		

8.2 Brake caliper adjustment screw

8.6 Remove the fixed pad split pin

8.7a The nylon washer on the floating pad

cover plate.

7   Clean the sprocket bore, and the spigot on the hub, before replacing the sprocket. Check the 'O' ring seal under the sprocket cover plate on the CB 550 models. Use new tab washers where fitted. Replace the other parts in reverse order. The chain adjusters should go back on the side from whence they came. CB 400F models: Grease and replace the spindle from the drive side.

8   Front brake: adjustment, checking the fluid and renewing the pads

1   The clearance between the fixed (inner) brake pad and the disc should be maintained at 0.15 mm (0.006 inch).
2   This may be adjusted by slackening the caliper adjuster locknut, and turning the adjuster until the inner (fixed) pad just touches the disc. Check this by rotating the wheel. Turn the adjuster back one half turn, and tighten the locknut. Rotate the wheel again to ensure that the pad does not touch the disc anywhere.
3   Inspect the pads and check that they are not worn below the incised red line running round the pad. If one or both is so worn, the pair **must** be renewed.
4   Remove the cap, washer and diaphragm from the master cylinder. Check that the fluid is up to the level line inside the reservoir. Check also that there are no small stones etc, in the reservoir; its exposed position invites vandalism.
5   To replace the pads, the caliper must be separated. Undo the two hexagon head bolts at the top of the caliper, and remove the inner caliper half with the fixed pad. Next, pull aside the outer caliper half. Do not put any strain on the rigid pipe, which it is not necessary to remove. If the rigid pipe is removed, or if **the front brake is operated** with the caliper dismantled, air will enter the hydraulic system.
6   Pull out the split pin from the fixed brake pad, and remove the pad. Take out the floating pad.
7   Note the nylon washer on the spigot inside the floating pad. When replacing the pads, they must locate on the dowels in both caliper halves. Renew the split pin. Apply silicone sealing grease to the sliding surfaces of the pads. This helps to keep out dirt and water. **Do not** get grease on the braking surface.
8   With the floating pad removed, the wheel cylinder operation may be checked. Apply the front brake very slightly, just enough to move the piston out a fraction, then push the piston back in by hand. It should move freely. No fluid should be apparent around the piston. Do not allow the piston to come right out!
9   Replace the caliper, and adjust as described in paragraph 2 of this Section.

8.7b Both pads have locating dowels

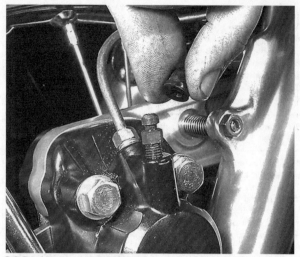

9.2 Remove the bleed nipple cap

9.4 Bleeding the front brake

10.10 The lower brake hose must be located in this clip

9 Front brake: bleeding

1   If brake action becomes spongy, it is likely that air has entered the hydraulic system, and must be bled out.

2   Remove the cap on the bleed valve nipple and replace it with a flexible tube. The other end of the tube should be submerged in a small quantity of clean hydraulic brake fluid in a jar.

3   Ensure that the reservoir is full to the level line. Replace the cap, to keep dirt out.

4   Pump the brake lever until full pressure is felt, and hold the lever against this pressure. Unscrew the bleed nipple about half a turn, and pull back on the lever until it touches the handlebar. Tighten the nipple.

5   Check the level of fluid in the reservoir, and repeat the operation until no more air bubbles emerge from the submerged end of the tube. Close the bleed valve nipple, remove the tube and replace the cap.

6   **Warning.** If fluid level in the reservoir is allowed to drop too low, air will re-enter the system and the whole routine will have to be repeated. The help of an assistant is recommended during this operation.

7   The fluid in the jar should not be re-used as it will have absorbed air. It must stand for at least 24 hours before all the air bubbles will disperse.

10 Hydraulic front brake system: examination and renovation

1   When the brake lever is squeezed, it moves a piston in the master cylinder so that a passage from the reservoir is blocked, and fluid is compressed in the master cylinder. This pressure is transmitted via the hydraulic line to the wheel cylinder. Here, the fluid pressure forces the piston outwards, which in turn presses the floating brake pad against the disc. Since the caliper is pivoted, reaction pulls the fixed pad against the disc, which is then clamped between both pads.

2   Remove the reservoir cap and diaphragm and check that fluid is up to the level line inside the reservoir. Top up only with hydraulic brake fluid of the correct grade; do not even mix brands. Make sure that the diaphragm, washer and cap are refitted properly.

3   **Never** allow fluid to come into contact with paintwork or plastic parts. Wipe off immediately in the event of spillage. It is a very effective paint stripper.

4   Brake fluid is hygroscopic, and the moisture that it absorbs will cause the fluid to become compressible. The system depends of course on the fluid being incompressible. But since the system has to be vented, it is inevitable that moisture will get in, and braking efficiency be impaired. Fluid suppliers suggest a useful life of one or two years for their product, but some authorities recommend renewal every six months. To do this, the old fluid must be pumped out completely via the bleed nipple. Cleanliness is essential, since grit will ruin the cylinders. Do not shake a tin of brake fluid, or leave it uncapped as this will entrap air.

5   Hoses, brake pipes, unions and the cylinders, should be checked regularly for signs of leakage. Also check the flexible hoses for splits or scuffing, and renew **immediately** if these are apparent. Make sure that neither the hoses nor the rigid pipe rub against any fixed part.

6   Check the security of the two caliper clamping bolts and the three bolts fixing the caliper pivot to the lower fork leg.

7   If any part of the system is dismantled, the system will require draining and refilling. To drain, attach a tube to the bleed nipple and put the other end in a container. Open the bleed nipple and pump the handlebar lever until all fluid is expelled. Do not re-use this fluid.

8   To refill, attach a tube to the bleed nipple and immerse the other end in some clean fluid. Fill the reservoir, open the bleed nipple and pump the lever until fluid issues from the end of the bleed tube. Do not allow the reservoir to empty completely, or

air will enter the system. Top up the reservoir and bleed as described previously.

9 The hydraulic line consists of a flexible hose from the master cylinder, to a junction fixed to the lower fork yoke. A hydraulic switch for the stop light is also screwed into this junction. A second hose leads from the junction to a rigid pipe, which is screwed into the caliper.

10 The upper hose is fixed by a banjo union at each end. There is a soft washer on each side of the banjo union joint. The lower hose has a banjo union fitting at the junction, and at the other end, a union nut on the rigid pipe screws into a fitting on the hose. The hose end fitting must be held with a spanner when turning the union nut.

11 When replacing the hydraulic hoses or pipe, make sure that they do not rub on any adjacent part. The rigid pipe must be routed in this way before tightening the union nut into the

caliper. Note the grommet on the lower hose, which locates in a clip on the centre mudguard bracket.

12 The hydraulic stop light switch can be removed, after draining the fluid, by pulling off the wires and unscrewing it. Note the soft washer on the stud. Do not overtighten the switch, when it is being replaced.

13 If the brake feels spongy and requires pumping, even after bleeding, it is possible that the hydraulic cylinders are faulty. Examine for signs of leaking fluid, which will indicate worn or damaged seals. If there is pressure at the lever, but the brake does not operate; or if the brake remains on, a piston has seized. See following section for cylinder overhaul.

14 There should be 2 - 5 mm (0.08 - 0.2 inch) free-play at the end of the handlebar lever. If there is more, the master cylinder should be examined and all worn parts renewed.

15 Check the disc thickness and flatness on a surface plate,

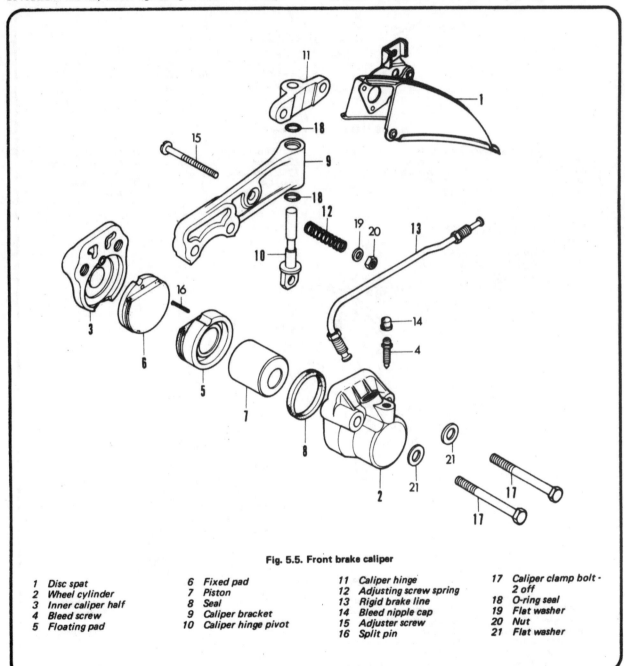

Fig. 5.5. Front brake caliper

1 Disc spat	6 Fixed pad	11 Caliper hinge	17 Caliper clamp bolt -
2 Wheel cylinder	7 Piston	12 Adjusting screw spring	2 off
3 Inner caliper half	8 Seal	13 Rigid brake line	18 O-ring seal
4 Bleed screw	9 Caliper bracket	14 Bleed nipple cap	19 Flat washer
5 Floating pad	10 Caliper hinge pivot	15 Adjuster screw	20 Nut
		16 Split pin	21 Flat washer

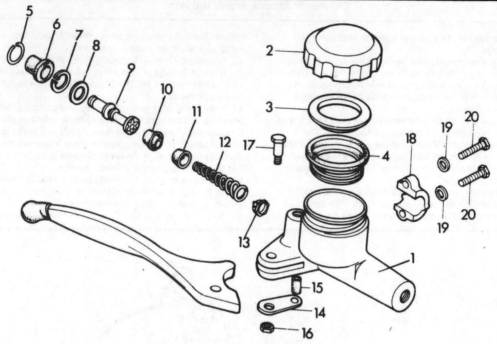

Fig. 5.6. Brake master cylinder and reservoir

1	Master cylinder and reservoir	6	Boot	11	Primary cup	16	Locknut
2	Reservoir cap	7	Circlip	12	Spring	17	Lever pivot screw
3	Washer	8	Washer	13	Check valve	18	Clamp
4	Diaphragm	9	Piston	14	Plate	19	Flat washer - 2 off
5	Wire circlip	10	Secondary cup	15	Bush	20	Bolt - 2 off

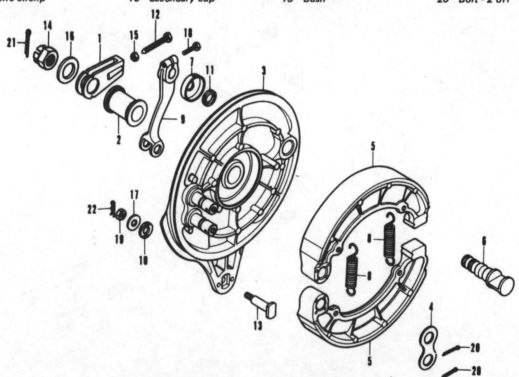

Fig. 5.7. Rear brake, CB 550 models (CB 400F is similar)

1	Chain adjuster - 2 off	8	Return spring - 2 off	14	Wheel spindle nut	20	Split pin - 2 off,
2	Spacer	9	Brake operating arm	15	Locknut - 2 off		CB 550 models
3	Brake plate	10	Rubber washer	16	Washer	21	Split pin
4	Brake shoe pivot plate	11	Seal	17	Washer	22	Spring pin for torque
5	Brake shoe - 2 off	12	Adjuster screw (chain) -	18	Brake operating arm		arm bolt
6	Brake operating cam		2 off		pinch bolt		
7	Wear indicator	13	Torque arm bolt	19	Nut		

Tyre changing sequence - tubed tyres

A Deflate tyre. After pushing tyre beads away from rim flanges push tyre bead into well of rim at point opposite valve. Insert tyre lever adjacent to valve and work bead over edge of rim.

B Use two levers to work bead over edge of rim. Note use of rim protectors

C Remove inner tube from tyre

D When first bead is clear, remove tyre as shown

E When fitting, partially inflate inner tube and insert in tyre

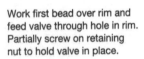

F Work first bead over rim and feed valve through hole in rim. Partially screw on retaining nut to hold valve in place.

G Check that inner tube is positioned correctly and work second bead over rim using tyre levers. Start at a point opposite valve.

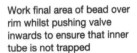

H Work final area of bead over rim whilst pushing valve inwards to ensure that inner tube is not trapped

after removing it from the wheel see Sections 3 and 4 of this
Chapter.

11 Hydraulic cylinders: examination and renovation

1   It is advisable to entrust this work to a specialist or dealer.
If either the master cylinder or the wheel cylinder requires
attention, the system must be drained, and the units removed
from the machine. Clinical cleanliness is essential when working
on the cylinders.
2   The master cylinder is integral with the reservoir. It is clamped
to the handlebar with two bolts. Remove the handlebar mirror
first. Unscrew the hose banjo union bolt, after pushing back the
cover. Unscrew the lever pivot nut and pivot screw and remove
the lever.
3   Remove the wire circlip and the boot from the cylinder,
taking care not to damage it. Remove the internal circlip. Push
or blow out the piston, primary cup, spring and check valve,
from the hose connection end. Take great care not to damage
any part, or score the cylinder. Note which way round the parts
are fitted.
4   Measure the cylinder and piston diameters. If either is scored,
or damaged by rust, it must be renewed.
5   Smear the bore of the cylinder with brake fluid, and re-install
the check valve followed by the spring (larger end first). Ensure
that the valve is correctly sealed. Smear a new primary cup with
fluid, and fit it into the cylinder. Refit the piston, circlip, boot
and wire retainer. Replace the lever and check the free-play.
6   Refit the master cylinder to the handlebar, connect up the
hose and fill and bleed the hydraulic system. Check brake
operation carefully.
7   The wheel cylinder is removed after draining and disconnecting
the rigid pipe at the caliper. Unscrew the two caliper bolts.
Remove the floating brake pad.
8   Tap the cylinder gently to extract the piston. Measure the
piston and cylinder diameters and renew if worn, scored or
damaged by rust.
9   Smear the cylinder bore and piston with brake fluid before
replacing the piston. Check that the piston moves freely. Replace
the brake pad.
10 Refit the wheel cylinder to the caliper assembly. Refill and
bleed the hydraulic system. Check pad adjustment and check
brake operation very carefully before using the machine.

12 Rear brake: adjustment

1   If there is excessive movement (more than 20 - 30 mm, ¾ -
1¼ inches) at the rear brake pedal, the brake requires adjustment.
If, with the brake fully on, the arrows of the brake wear
indicator align, the linings should be inspected for wear (not
fitted to CB 550).
2   With the bike on its centre stand, and the rear wheel clear of
the ground, check that the brake is not binding.
3   Turn the brake rod adjuster clockwise until the brake is fully
on, then slacken the adjuster a few turns until there is no
binding.
4   The brake pedal height may be adjusted by the bolt under the
right-hand footrest on the 400, or under the swinging arm pivot
nut on the 550. The pedal should be just under the ball of the
foot when seated normally.
5   Check the rear stop light operation if either the brake has
been adjusted, or the pedal height has been changed. The stop
light switch is adjusted by turning the hexagonal flanged bush.
6   Make sure the brake torque arm bolts are tight. A serious
accident can result from the torque arm coming loose.

13 Rear brake: examination and renovation

1   Except on the CB 550 model, the condition of the brake
linings may be assessed by means of the wear indicator on the

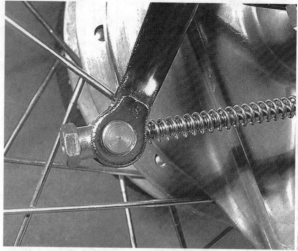
12.3 Turn hexagonal adjuster to adjust rear brake

13.2 Remove the brake plate

13.5 Brake plate of CB 400F model

brake shoe pivot. When the brake is applied fully, if the two arrows align, the linings should be examined for wear.

2 Remove the rear wheel as described previously. On the CB 400F model, remove the spindle nut, washer, chain adjuster and spacer from the brake side. Take off the brake plate.

3 Braking lining thickness may be checked without removing the shoes from the back plate. Renew the shoes if the linings are less than the recommended thickness (2 mm, 0.08 inches).

4 Both the CB 400F and CB 550 models have a single leading shoe rear brake. On the CB 400F model, the shoes have semi circular ends which pivot in semi circular housings, integral with the brake back plate. The CB 550 models have conventional pivoted shoes, with a pivot pin for each.

5 To remove the brake shoes on the CB 400F model, remove the split pin and washer at the shoe pivot. Pull the shoes apart, and upwards into a vee formation. The shoes can then be removed from the back plate and the spring unhooked.

6 On the CB 550 models, remove the two split pins from each shoe pivot pin, then the pivot pin plate. Unhook both return springs, pull the shoes apart and slide them off their pivots.

7 Badly soiled linings must be renewed, but slight oil or grease contamination may be removed by scrubbing in detergent. Remove glaze by wire brushing. The leading ends of the linings should be chamfered to avoid 'grab'.

8 Clean and grease the pivot pins or housings, and the cam. Do not overgrease or it will run and contaminate the brake linings.

9 The cam spindle should be greased by removing the cam. Remove the brake arm pinch bolt, and pull off the arm, wear indicator ('F' models only), washer (on the CB 400F model), and seal. Note the location dots on the brake arm and cam spindle also the position of the wear indicator and replace them in the same position.

10 Blow out all dust from the brake drum, and examine the braking surface for scores. Shallow scores may be removed by skimming the drum in a lathe. This must be done with the hub laced into the rim and is a specialised operation. Clean the drum with a petrol soaked rag.

11 Replace all components in reverse order. Renew the split pins. Also renew the brake shoe return springs if they are weakened. Compare them with the length of new springs.

14 Rear wheel sprocket and cush drive: examination and renovation

1 Remove the rear wheel as described in Section 6 of this Chapter.

2 Examine the teeth on the sprocket. If they are chipped or hooked, the sprocket should be renewed. Sprocket removal is described in Section 7.3 of this Chapter. On the CB 400F model, the studs for the cush drive are bolted to the sprocket and should be removed. The cover plate of the CB 400F model is rivetted on and will be supplied with the new sprocket.

3 Check the circumferential play of the sprocket when it is assembled on the hub. It should be negligible. If not, first check the security of the sprocket fixing nuts, then dismantle the cush drive.

4 The cush drive on the CB 400F model consists of the four studs already mentioned, which are a tight fit in rubber bushed holes in the hub. The rubber bushes are unlikely to require renewal.

5 The CB 550 models employ a vane type cush drive. It is not necessary to unbolt the sprocket as it will tap off together, with the drive flange. Note the 'O' ring seal on the hub spigot. There are four pairs of damper elements, each of which are pegged into the hub. The larger one leads, in the direction of rotation.

6 Clean the spigot on the hub and the sprocket bore, before replacing the sprocket. Replace all parts in reverse order.

7 Sprocket sizes have been selected by the manufacturer after exhaustive testing and no advantage will be obtained by alteration (other than when using a sidecar).

15 Rear chain: examination and renovation

1 The rear drive chain operates under arduous conditions, unprotected exposed to water and grit, and unlubricated. Wear under these conditions can be rapid and chain tension should be checked regularly.

2 A slack or worn chain may jump the sprockets and will certainly accelerate wear of the sprocket teeth. An overtight chain will impose excessive strain on the gearbox and rearwheel, shortening life of the components and absorbing power.

3 The chain is correctly adjusted when, with the machine on the centre stand, there is 20 mm (¾ inch).

4 The chain must be lubricated regularly, especially in wet weather. This may be done in various ways, the most convenient and effective being to use one of the aerosol chain lubricants.

5 As an alternative, the chain can be removed completely, and washed in paraffin and lubricated in a grease such as Linklyfe, at longer intervals. To remove the chain, find the link (when fitted), unclip the spring clip using flat pliers, and take out the link. To avoid removing the gearbox cover, link an old chain to the one fitted and draw it over the gearbox sprocket. When an endless

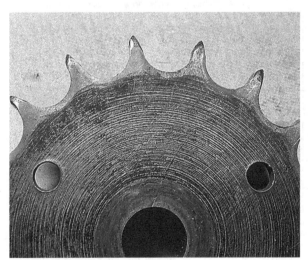

14.2 Badly hooked sprocket teeth

16.2 Rear wheel alignment marks

chain is fitted, the rear wheel and gearbox end cover must be removed, so that the chain may be unhooked.

6   Whilst the chain is removed, and after it has been washed, examine it for wear. Anchor one end, compress the chain and measure its length. Then pull the chain out straight, and measure the increased length. If this exceeds 6 mm per 300 mm (¼ inch per foot), the chain is due for renewal. Also check for damaged rollers or loose or broken side plates.

7   Whilst the chain is off, examine the sprockets for hooking or chipping of the teeth. Renew if either of these faults are present. Remove the left side gearbox cover to examine the gearbox sprocket. It is preferable to renew both sprockets and the chain is unison, so that the rate of wear is not accelerated by running old and new parts together.

8   To replace a chain fitted with a spring link, the gearbox cover must be removed, unless a spare chain has previously been drawn over the gearbox sprocket. Always fit the spring clip with its closed end facing direction of travel of the chain. Failure to do so could result in the clip being knocked off. The easiest way to refit the link is to position each end of the chain in adjacent teeth on the rear sprocket, then insert the link.

9   In the event of chain breakage, an extra link or links may be inserted to effect a temporary repair. Remove rivets with a chain rivet extractor. Renew the chain at the earliest opportunity.

16 Wheel alignment

1   It is important that the front and rear wheels are maintained in correct alignment, see accompanying diagram. Incorrect alignment will affect handling, and accelerate type, chain and sprocket wear.

2   Although the index marks on the rear forks give a quick visual check that the wheels are in line, the alignment should be checked more accurately at regular intervals.

3   The easiest way is to sight down the wheels by eye, just below wheel spindle height. More accurately, use a stretched cord or long straight plank alongside the machine at wheel spindle height. Measure from this datum to two opposite points on each rim (don't use the tyres, as they are of different section). The measurements should all be equal, assuming the wheels are in correct alignment.

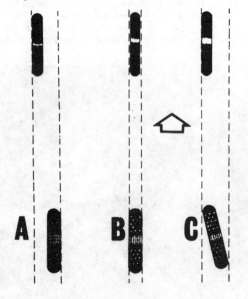

Fig. 5.8. Wheel alignment

a   Wrong   b   Correct   c   Wrong

17 Tyres: removing and replacing

1   At some time or other the need will arise to remove and replace the tyres, either as the result of a puncture or because a renewal is required to offset wear. To the inexperienced, tyre changing represents a formidable task, yet if a few simple rules are observed and the technique learned, the whole operation is surprisingly simple.

2   To remove the tyre from either wheel, first detach the wheel from the machine by following the procedure in Sections 3 or 6, depending on whether the front or the rear wheel is involved. Deflate the tyre by removing the valve insert and when it is fully deflated, push the bead of the tyre away from the wheel rim on both sides so that the bead enters the centre well of the rim. Remove the locking cap and push the tyre valve into the tyre itself.

3   Insert a tyre lever close to the valve and lever the edge of the tyre over the outside of the wheel rim. Very little force should be necessary; if resistance is encountered it is probably due to the fact that the tyre beads have not entered the well of the wheel rim all the way round the tyre.

4   Once the tyre has been edged over the wheel rim, it is easy to work around the wheel rim so that the tyre is completely free on one side. At this stage, the inner tube can be removed.

5   Working from the other side of the wheel, ease the other edge of the tyre over the outside of the wheel rim which is furthest away. Continue to work around the rim until the tyre is free completely from the rim.

6   If a puncture has necessitated the removal of the tyre, re-inflate the inner tube and immerse it in a bowl of water to trace the source of the leak. Mark its position and deflate the tube. Dry the tube and clean the area around the puncture with a petrol soaked rag. When the surface has dried, apply the rubber solution and allow this to dry before removing the backing from the patch and apply the patch to the surface.

7   It is best to use a patch of the self-vulcanising type which will form a very permanent repair. Note that it may be necessary to remove a protective covering from the top surface of the patch, after it has sealed in position. Inner tubes made from synthetic rubber may require a special type of patch and adhesive if a satisfactory bond is to be achieved.

8   Before refitting the tyre, check the inside to make sure that the object which caused the puncture is not trapped. Check the outside of the tyre, particularly the tread area, to make sure nothing is trapped that may cause a further puncture.

9   If the inner tube has been patched on a number of past occasions, or if there is a tear or large hole, it is preferable to discard it and fit a new one. Sudden deflation may cause an accident, particularly if it occurs with the front wheel.

10 To replace the tyre, inflate the inner tube sufficiently for it to assume a circular shape but only just. Then push it into the tyre so that it is enclosed completely. Lay the tyre on the wheel at an angle and insert the valve through the rim tape and the hole in the wheel rim. Attach the locking ring on the first few threads, sufficient to hold the valve captive in its correct location.

11 If the tyre has a balance mark (usually a dot), as on the tyres fitted as original equipment, this must be positioned alongside the valve. Similarly, any arrow indicating direction of rotation must face the right way.

12 Starting at the point furthest from the valve, push the tyre bead over the edge of the wheel rim until it is located in the central well. Continue to work around the tyre in this fashion until the whole of one side of the tyre is on the rim. It may be necessary to use a tyre lever during the final stages.

13 Make sure that there is no pull on the tyre valve and again commencing with the area furthest from the valve, ease the other bead of the tyre over the edge of the rim. Finish with the area close to the valve, pushing the valve up into the tyre until the locking cap touches the rim. This will ensure the inner tube is not trapped when the last section of the bead is edged over

the rim with a tyre lever.

14 Check that the inner tube is not trapped at any point. Re-inflate the inner tube and check that the tyre is seating correctly around the wheel rim. There should be a thin rib moulded around the wall of the tyre on both sides which should be equidistant from the wheel rim at all points. If the tyre is unevenly located on the rim, try bouncing the wheel when the tyre is at the recommended pressure. It is probable that one of the beads has not pulled clear of the centre well.

15 Always run the tyres at the recommended pressures and never under or over-inflate. The correct pressures for solo use are given in the Specifications Section of this Chapter. If a pillion passenger is carried, increase the rear tyre pressure only by approximately 4 psi.

16 Tyre replacement is aided by dusting the side walls, particularly in the vicinity of the beads, with a liberal coating of French chalk. Washing up liquid can also be used to good effect, but this has the disadvantage of causing the inner surfaces of the wheel rim to rust.

17 Never replace the inner tube and tyre without the rim tape in position. If this precaution is overlooked there is good chance of the ends of the spoke nipples chafing the inner tube and causing a crop of punctures.

18 Never fit a tyre which has a damaged tread or side walls. Apart from the legal aspects, there is a very great risk of blow-out, which can have serious consequences on any two-wheel vehicle.

19 Tyre valves rarely give trouble, but it is always advisable to check whether the valve itself is leaking before removing the tyre. Do not forget to fit the dust cap, which forms an effective second seal. This is especially important on high performance machines where centrifugal force can cause the valve insert to retract and the tyre to deflate without warning.

18 Valve cores and caps

1 Valve cores seldom give trouble, but do not last indefinitely.

Dirt under the seating will cause a puzzling 'slow puncture'. Check that they are not leaking by applying spittle to the end of the valve, and watching for air bubbles.

2 A valve cap is a safety device, and should always be fitted. Apart from keeping dirt out of the valve, it provides a second seal in case of valve failure, and may prevent an accident resulting from sudden deflation.

19 Front wheel balancing

1 The front wheel should be statically balanced, complete with tyre. An out of balance wheel can produce dangerous wobbling at high speed.

2 Some tyres, such as those fitted as original equipment to the CB 400F and CB 550F models, have balance mark on the sidewall. This must be positioned adjacent to the valve. Even so, the wheel still requires balancing.

3 With the front wheel clear of the ground, spin the wheel several times. Each time, it will probably come to rest in the same position. Balance weights should be attach diametrically opposite the heavy spot, until the wheel will not come to rest in any set position, when spun.

4 Balance weights, which clip round the spokes, are available in 5, 10 or 20 gramme weight. If they are not available, wire solder wrapped round the spokes and secured with insulating tape will make a substitute.

5 It is possible to have a wheel dynamically balanced at some dealers. This requires its removal.

6 There is no need to balance the rear wheel under normal road conditions, although any tyre balance mark should be aligned with the valve.

20 Fault diagnosis: wheels, brakes and tyres

Symptom	Cause	Remedy
Handlebars oscillate at low speeds	Buckle or flat in rim, most probably front wheel	Check rim alignment. Correct by retensioning spokes or renewing rim.
	Tyre not straight on rim	Check tyre fitting.
Machine lacks power and accelerates poorly	Brakes binding	Warm drum or disc provides best evidence. Re-adjust rear brake or check operation of front brake.
Rear brake grabs when applied	Ends of linings not chamfered	Chamfer with file.
	Drum elliptical	Skim brake drum (specialist operation).
Front brake feels spongy	Air in hydraulic system	Bleed system.
Front brake efficiency poor or brake remains on	Seized cylinder	Check brake operation, free cylinder if necessary.
Rear brake pull-off sluggish	Brake cam pivot binding	Free and grease.
	Weak shoe return springs	Renew springs.
Harsh transmission	Worn or maladjusted rear chain	Adjust or renew.
	Worn sprockets	Renew.
	Worn or deteriorating cush drive rubbers	Renew.
Heavy steering	Low front tyre pressure	Check.
Front and rear wheel wobble	Worn bearings	Check and renew if necessary.

Chapter 6 Electrical system

Contents

Specifications

	CB400F	CB550	CB550F
Alternator	Three-phase field-excited type		
Output	0.156 kw @ 5000 rpm	0.11 kw @ 2000 rpm	0.11 kw @ 2000 rpm
Stator coil resistance ...	0.61 - 0.69 ohms	0.35 ohms ± 10%	0.35 ohms ± 10%
Field coil resistance ...	4.6 - 5.0 ohms	4.9 ohms ± 10%	4.9 ohms ± 10%
Battery	12V, 12 Ah	12V, 12 Ah	12V, 12 Ah
Polarity	Negative earth		
Regulator	Electro-mechanical or electronic		
Core gap	0.6 - 1.0 mm (0.020 - 0.040 in.)	0.6 - 1.0 mm (0.020 - 0.040 in.)	0.6 - 1.0 mm (0.020 - 0.040 in.)
Points gap	0.3 mm (0.012 in.)	0.2 mm (0.008 in.)	0.2 mm (0.008 in.)
Starter	Drip-proof d.c. motor, driving through centrifugal clutch		
Motor	12V, 0.6 kw	12V, 0.6 kw	12V, 0.6 kw
Minimum brush length ...	5.5 mm (0.22 in.)	—	—
Brush spring festoon ...	0.4 kg (·8 lb)	—	—
Bulbs	(All are 12 volt)		
Headlamp (sealed beam in US)	35/50W (US), 50/40W or 35/35W (UK)	40/50W	40/50W
Tail/stop lamp	8/23W SCC index	4/32W	3/32 cp SCC index
Direction indicators			
Front	8/23W (US), 23 or 18W (UK) SCC	32W	8/23W SCC index
Rear	25 or 18W SCC	32W	25W SCC
Pilot lamp (UK only) ...	3.4W MCC 8.8 mm tubular	—	3.4W MCC 8.8 mm tubular
Instrument lighting ...)	3.4W (4)	2W	3.4W (4)
Neutral indicator ...)	3.4W	2W	3.4W
Direction indicator warning ... ...)11 MCC	3.4W (2)	2W	3.4W (2)
Main beam warning ...)mm	3.4W	2W	3.4W
Oil pressure warning ...) round	3.4W	2W	3.4W
Fuses			
Size	32 mm long X 6.5 mm diameter (1¼ inch long X ¼ inch diameter)		
Main	15A	15A	15A
Headlamp	7A	—	7A
Tail lamp	7A	7A	—

1 General description

A 12 volt negative earth system is fitted to all the models covered by this manual.

The crankshaft mounted alternator charges the battery through a silicon bridge rectifier.

The alternator is of the three phase, field excited, brushless type. The output is controlled by altering the field current, using an electro-mechanical or electronic regulator. The ac output from the generator has to be converted to dc for charging the battery. This is accomplished by the silicon rectifier.

All the electrical components are fitted beneath the dualseat.

CB 400F and CB 500F models imported to the USA differ from UK models, and also the earlier CB 550 models, in having a headlamp which is switched on with the ignition. Also, a double filament lamp is fitted in the front flashers, which lights when the main headlamp beam is selected.

The solenoid operated starter is a dc motor, driving the crankshaft through a centrifugal clutch and gears on the gearbox input shaft. The solenoid switch is a magnetic device capable of switching the high current required by the motor. It is mounted behind the frame side cover, on the left.

A switch operated by the clutch lever, in conjunction with the neutral switch and a diode, prevents operation of the starter motor when in gear with the clutch engaged; this applies to CB 400F and CB 550F models only. The engine stop switch on the right-hand switch console both cuts out the ignition, and prevents operation of the starter.

2 Battery: examination, topping-up and maintenance

1 The battery is housed behind the right-hand side cover, and is fixed with a rubber strap.

2 The transparent plastics case of the battery permits the upper and lower levels of the electrolyte to be observed without disturbing the battery by removing the side cover. Maintenance is normally limited to keeping the electrolyte level between the prescribed upper and lower limits and making sure that the vent tube is not blocked. The lead plates and their separators are also visible through the transparent case, a further guide to the general condition of the battery.

If electrolyte level drops rapidly, suspect over-charging, and check the system.

3 Unless acid is spilt, as may occur if the machine falls over, the electrolyte should always be topped up with distilled water to restore the correct level. If acid is spilt onto any part of the machine, it should be neutralised with an alkali such as washing soda or baking powder and washed away with plenty of water, otherwise serious corrosion will occur. Top up with sulphuric acid of the correct specific gravity (1.260 to 1.280) only when spillage has occurred. Check that the vent pipe is well clear of the frame or any of the other cycle parts.

4 To remove the battery, first take off the frame cover on the right. Disconnect the battery leads, and vent pipe. Unhook the rubber retaining strap, and lift out the battery. Make sure that all the rubber battery pads are in position.

5 It is seldom practicable to repair a cracked battery case because the acid present in the joint will prevent the formation of an effective seal. It is always best to renew a cracked battery, especially in view of the corrosion which will be caused if the acid continues to leak.

6 If the machine is not used for a period, it is advisable to remove the battery and give it a 'refresher' charge every six weeks or so from a battery charger. The battery will require recharging when the specific gravity falls below 1.2 per cell (at 20°C - 68°F). The hydrometer reading should be taken at the top of the meniscus with the hydrometer vertical. If the battery is left discharged for too long, the plates will sulphate. This is a grey deposit which will appear on the surface of the plates, and will inhibit recharging. If there is a sediment on the bottom of the battery case, which touches the plates, the battery needs

to be renewed. Limit charging rate to 2A. If charging from an external source with the battery on the machine, **disconnect the leads, or the rectifier will be damaged.** Keep naked flames away from the filler vents.

7 Make sure that the vent tube is routed correctly (our new from the dealer's one wasn't - in spite of the label on the frame!).

8 Occasionally, check the condition of the battery terminals to ensure that corrosion is not taking place, and that the electrical connections are tight. If corrosion has occurred, it should be cleaned away by scraping with a knife and then using emery cloth to remove the final traces. Remake the electrical connections whilst the joint is still clean, then smear the assembly with petroleum jelly (NOT grease) to prevent recurrence of the corrosion. Badly corroded connections can have a high electrical resistance and may give the impression of complete battery failure.

3 Electrical system : general

1 The generator output can be checked by using either an individual ammeter and dc voltmeter, or a multimeter set to the appropriate scale, as described below. Prior to making any tests on the electrical system ensure that the battery is in good condition and fully charged.

2 Disconnect the positive battery lead and connect the positive (+) voltmeter lead to the positive battery terminal and the negative (−) voltmeter lead to a good earth point. Connect the ammeter positive (+) lead to the battery positive lead and the ammeter negative (−) lead to the battery positive terminal. Start the engine and compare the meter readings with those shown below at the given engine speeds. Remember to check the amperage reading with the lighting switch in the 'On' and 'Off' positions.

3 Slight irregularities in the test results can be corrected by adjusting the voltage regulator, but any significant differences should be referred to an authorized Honda agent or auto-electrical specialist for attention. In the event of low or no output from the generator check the wiring connections and the generator coil resistances.

CB400 F model

Engine rpm	1000	2000	3000	4000
Charging current:				
Nightime running	1.6	1.9	2.0	1.8
Daytime running	−	−	4	2.6
Battery terminal volt.	12.5	14.2	15	15

Engine rpm	5000	6000	7000	8000
Charging current:				
Nightime running	1.6	1.5	1.4	1.4
Daytime running	2.0	1.6	1.4	1.4
Battery terminal volt.	15	15	15	15

CB550 models

Engine rpm	1000	2000	3000	4000
Charging current:				
Nightime running	2-3	1	1	1
Daytime running	6.5	−	2.4	1.3
Battery terminal volt.	12	12.4	13.2	14.5

Engine rpm	5000	6000	7000	8000
Charging current:				
Nightime running	1	1	1	1
Daytime running	1	1	0.8	0.6
Battery terminal volt.	14.5	14.5	14.5	14.5

4 Alternator: examination and renovation

1 The alternator is located on the left-hand end of the crankshaft. It consists of a rotor, which surrounds a field coil, and is in turn surrounded by the generator coils. The field coils magnetises the rotor by induction. There are no brushes or slip rings.

2   Remove the left-hand gearbox cover, and unplug the generator wires (green, white and three yellow). Remove the three screws, and the alternator cover. The rotor is bolted to the end of the crankshaft, and will not require attention. The field and generator coils are each screwed to the cover with three screws. It is not necessary to remove any coil unless renewal is required.

3   Test the resistance of each coil. The field coil should measure 4.9 ohms ± 10% between the white and green leads. The resistance between any two of the three yellow generator coil leads in turn should be 0.61 - 0.69 ohms for the CB 400F model or 0.35 ohms ± 10% for the CB 550 models.

4   The outer generator coil fixing screws are accessible from inside the cover. The inner field coil fixing screws are accessible after unscrewing two screws and removing the 'Honda' name plate on the CB 400F model, or after prising off the name plate of the CB 550 models.

5   Rectifier: examination and renovation

1   The silicon diode bridge rectifier is located between the side frame covers, to the rear of the battery on the CB 400F model or beneath the indicator relay of the CB 550 models. It is secured by a single hexagonal nut.

2   Each diode should be checked for forward and reverse continuity with an ohmmeter or dry battery and bulb test circuit. Unplug the rectifier wiring connector. With one probe of the test instrument placed on one of the yellow lead pins, connect the other to the red-with-white and green pins, in turn. Then replace this check using each of the other two yellow lead pins. Reverse the polarity of the probes and repeat the tests again. If there is no continuity in either or both directions on any one diode, the rectifier must be renewed. Do not use an ohmmeter's 10 to the power of 6 ohms range for this test.

3   **Do NOT** run the engine with the red-with-white rectifier lead disconnected. Be sure to disconnect the rectifier plug when charging the battery from an external source. Failure to observe either of these two precautions may result in permanent damage to the rectifier.

6   Voltage regulator: examination

1   Intermittent opening of the regulator contacts creates a resistance in the alternaor field circuit, so reducing output. The voltage at which this opening occurs may be adjusted. Some models are fitted with a non-adjustable solid state regulator. The screws securing the cover of the regulator are sealed, and tampering will no doubt nullify the warranty. If faulty regulation is suspect, as indicated by a flat or an overcharged battery, refer to a Honda dealer.

2   The regulator is mounted in the frame under the left-hand side cover. It is fixed by two screws in rubber bushes. A colour coded strip on the cover indicates the colours of the wires to the regulator. Connect the wire to the terminal adjacent to the corresponding colour on the strip.

3   If the charge rate does not conform to the specifications given in Section 3 the regulator should be checked. Remove the regulator cover and locate the voltage adjusting screw positioned at the end of the unit and secured by a locknut. Slacken the locknut and turn the screw clockwise to increase the charge rate or anticlockwise to decrease it. Make only a small adjustment at a time and check the setting as described in Section 3. Tighten the locknut on completion of adjustment.

4   Note that there will be a 0.5 volt rise in voltage as the low speed contacts change over to the high speed contacts in the regulator unit. If the change in voltage is greater than 0.5 volts or if there is a drop in voltage, the core gap probably requires adjustment.

5   With the ignition turned off, inspect the condition of the electrical contact faces. If they are dirty or pitted they can be cleaned with fine emery paper. Check the core gap using a feeler gap. If the gap does not fall within the figure given in the

Specifications at the beginning of this Chapter, slacken the adjusting screw (located at the opposite end of the unit to the voltage adjusting screw) and move the point body to obtain the required clearance.

6   It is advisable to check the condition of the other set of points and if necessary adjust to the figure given in the Specifications. To adjust the point gap slacken the screw located on the side of the lower point and reposition the lower point accordingly.

7   Refit the regulator cover and check the charge rate once again. If it is not possible to obtain the desired results the regulator should be renewed or checked by an authorized Honda agent.

7   Wiring: general

1   The wiring is colour coded in accordance with the accompanying wiring diagrams. There are slight differences between USA and UK models ie; USA models have no pilot light, and the front direction indicators light when the headlamp main beam is on.

2   The wiring should be examined for damaged insulation, poor connections and particularly poor earth connections. The wires should not be pulled tight, nor have tight bends. They should not be routed over sharp edges.

3   Check the continuity of the wires to eliminate this possible cause of failure. In the following continuity tests, the item being tested should be disconnected unless otherwise stated. Always disconnect the battery earth (negative) lead before working on the electrical system. On the CB 400F and CB 550F models, these connections are inside a cover beneath the petrol tank, on the left-hand side of the frame spine. On the CB 550 model, all connections from the handlebar switches are made inside the headlamp shell.

4   Most electrical components and connectors are located behind the frame cover on the left-hand side, with the exception of the rectifier on the CB 400F model which is to the rear of the battery.

8   Main switch: examination

1   On CB 400F and CB 550F models, the main switch is secured to the top fork yoke with two hexagonal bolts. For USA models it also operates the steering lock. Connections to the switch are

7.4 Electrical compartment behind left-hand side cover (CB 400F) 1. Regulator: 2. Flasher relay; 3. Fuse holder 4. Starter solenoid; 5. Clutch cut-out diode; 6; Solenoid wiring connector

made via the integral five pin plug. Disconnect the battery earth lead, unplug the connector and check as described above.

2   The switch on the CB 550 model is mounted on the frame, just below the front of the petrol tank, on the left. Disconnect the battery earth (negative) lead, and pull apart the switch connector. There should be no continuity with the switch off. With the switch on, check continuity in accordance with the table on the wiring diagram; the connector socket layout is shown in Fig. 6.2.

9 Lighting switch: examination

1   This section applies to the CB 550 model and UK CB 400F model only. On the CB 550 model this switch also controls the main and dip headlamp beams.

2   Switch off, and disconnect the brown with blue, black and black with red, wires on the CB 400F model, or the black, blue, brown with white and white, wires on the CB 550 model. Check continuity in accordance with the table on the wiring diagram. Continuity is indicated by the line connecting the circles thus:
0————— 0

10 Headlamp dip switch: examination

1   There are differences in the switching arrangements. On the UK CB 400F models, a switch on the right controls all lights with a dipswitch on the left. On the CB 550 model a switch on the right controls the pilot, head and dip beams (see Section 8), as there is no separate dip switch. On the CB 550F model and USA CB 400F model the headlamp is always on when the ignition is switched on, and only a dip switch is provided.

2   Switch off the main switch and disconnect the black with yellow, blue and white wires. Check continuity in accordance with the table on the wiring diagram.

11 Direction indicator switch: examination

1   If the indicators malfunction, the switch should be checked for continuity. Switch off, and disconnect the grey, orange and light blue wires. Check for continuity between grey and orange for left, and grey and light blue for right switch positions.

12 Horn and horn button: examination

1   The horn is provided with an adjusting screw in the back of the horn body so that the volume can be varied, if necessary. To adjust the horn note, turn the screw not more than one half turn in either direction and test. If the note is weaker, or lost altogether, turn in the opposite direction. Continue adjusting by one half turn at a time until the desired volume and note is obtained.

2   The horn may be checked for continuity if it fails. Switch off, and disconnect the wires to the horn (light green and black). There should be continuity through the horn. Alternatively, check the horn with a fully charged 12v battery.

3   The horn button may also be checked for continuity when depressed. Switch off and disconnect the light green and the black wires on the UK CB 400F models; or light grey only on USA CB 400F models and the 550's. Check for continuity between the two wires on UK CB 400F models or between light green and frame on USA CB 400F models and the 550's.

4   The headlamp flash, on UK CB 400F models operated by sliding the horn button, may be tested for continuity between the black and the black with yellow wires when operated as described above.

13 Neutral indicator switch: examination

1   A spring loaded plunger on the CB 400F model is grounded through a contact on the selector drum. On the CB 550 models, a spring blade contact is grounded via a cam on the gear selector drum, to operate the neutral warning light. Both switches are situated behind the left-hand gearbox cover above the oil pump.

2   Switch off and disconnect the wires to the switch (light green with red). Check continuity between the terminal and frame when in neutral on all models. On the CB 550 models the blade may be bent to complete contact.

14 Oil pressure switch: examination

1   The oil pressure switch should operate when oil pressure drops below 0.5 kg/sq. cm (7 lb/sq. in) on the 550's, or 0.3 kg/sq. cm (4.3 kb/sq. in) on the 400. If the warning lamp comes on and remains on, stop the engine immediately and find the reason. Check the oil level in the sump. If correct and there is no reason to suspect lubrication system failure, check the warning lamp circuit.

2   There should be continuity across the switch only when the engine is not running. Disconnect the wires and check. The switch is screwed into the top of the oil pump. Renew the switch to make a positive check of switch failure.

15 Stop lamp switches: examination and adjustment

1   The front brake stop lamp switch is hydraulic and non-adjustable. It is screwed into the hydraulic junction on the lower fork yoke. Switch off, disconnect the wires at the switch, and check for continuity when the front brake is operated.

2   The rear brake stop lamp switch is mounted on the right-hand frame down tube. Switch off, disconnect the wires at the line connectors and check for contiguity when the brake is operated.

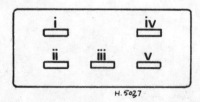

H. 5027

Fig. 6.1. Illustration showing front view of main switch plug (CB 400F and CB 550F models)

(i)   Brown
(ii)   Brown with white
(iii) Black
(iv)   Red
(v)   Brown

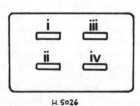

H.5026

Fig. 6.2. Illustration showing front view of main switch socket (CB 550 model)

(i)   Black
(ii)   Brown with white
(iii) Brown
(iv)   Red

3   Adjust the rear stop light switch by turning the hexagonal threaded bush. If it is raised, the stop lamp will operate earlier, and vice versa.

16 Handlebar switches: removing

1   If the handlebar switches require removal for any reason, first disconnect all wires at the terminals in the terminal cover beneath the tank (CB 400F and 550F models) or headlamp shell (CB 550 model) after switching off. The switch housings are split, and are clamped together on the handlebar by screws in the underside. The switch housing on the right also houses the twist grip drum. The throttle cables will have to be disconnected from the drum, to remove the bottom half of the housing.

17 Starter motor: examination and renovation

1   The starter motor is mounted on the crankcase, behind cylinder number one. It drives the crankshaft through a centrifugal clutch on the gearbox input shaft.
2   Before removing the starter motor, disconnect the starter cable from the solenoid. Remove the left-hand side frame cover, push aside the solenoid rubber cover and undo the terminal nut. Remove the ring terminal and replace the nut and washers.
3   On the CB 550 models only, unscrew the two hexagon head screws and remove the starter motor cover on the top of the gearbox. On all models, remove the left-hand crankcase cover. Unscrew the two hexagon head starter fixing bolts. Pull the starter motor out towards the left, it will be fairly tight. Examine the 'O' ring seal on the spigot which locates in the crankcase.
4   Unscrew the long starter motor clamping screws, and remove the commutator end cover (that's the end opposite the drive pinion). Take care not to lose the shims on the armature shaft. Examine the commutator, brushes and brush springs. The commutator should be clean, smooth and unworn. The brushes should not be shorter than 5.5 mm (0.22 inch). Spring tension should not be less than 0.4 kg (0.8 lb). Check also the condition of the commutator end cover 'O' ring seal.
5   The brush tails are secured with screw terminals; lift each spring and withdraw the brush from its holder. Unscrew the terminal nut and the brush may be removed for renewal. Note that the insulated brush tail is connected to the field coil terminal.

15.3 Rear brake stop lamp switch is adjustable

17.2 Disconnect the starter cable. Note also the diode (arrowed)

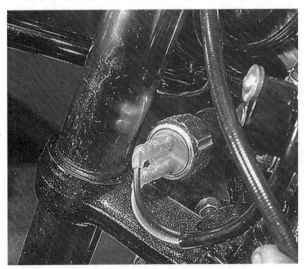

15.1 Front brake stop lamp switch on lower fork yoke

17.3a Unscrew the starter fixing bolts ...

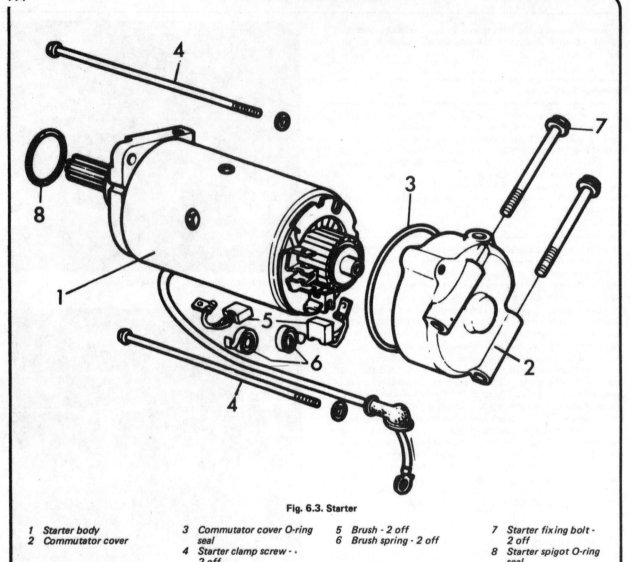

Fig. 6.3. Starter

1 Starter body	3 Commutator cover O-ring seal	5 Brush - 2 off	7 Starter fixing bolt - 2 off
2 Commutator cover	4 Starter clamp screw - 2 off	6 Brush spring - 2 off	8 Starter spigot O-ring seal

17.3b ... and pull out the starter

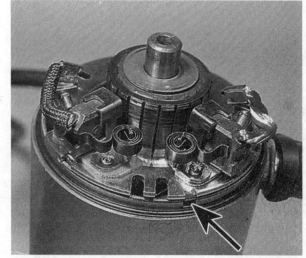

17.7 Wedge the brushes in raised position. Note also the locating notch in the brush holder plate

6 A dirty commutator will impair starter efficiency. Clean with fine glass paper, **never** emery cloth, and wash. The commutator segments must be undercut; that is, the insulation between each copper segment should be below the level of the segment. The easiest way to undercut is with a hacksaw blade, ground to the thickness of the insulation. Undercut approx. 0.3 mm (0.012 inch). If the commutator is worn or damaged, it may be skimmed by a specialist, provided that the finished diameter is not too small. When removing the brush holder plate, which is held only by the field coil terminal, note that it locates in a notch in the starter body.

7 If the motor fails to turn when the starter button is pressed, and the brushes and commutator are in good condition, check the field coil and armature coils. Raise the brushes in their holders and wedge them in place with their springs. Check that there is continuity between the starter cable and the brush connected to the field coil. Check for continuity between adjacent commutator segments and that there is no continuity between the commutator and the armature spindle.

8 Before replacing the starter motor, examine the starter pinion for wear or damage and check the security of the starter cable terminal nut. The commutator end cover locates on a tag on the brush holder plate and a line stamped on the starter aligns with a raised line on the cover.

18 Starter solenoid switch: examination and renovation

1 The starter motor requires a large current, approaching 100 amps to operate. To minimise resistance, a large cable is needed to carry this current and a switch with heavy duty contacts. To enable the handlebar push switch to fulfill these requirements, a remotely located magnetic switch is used. The starter button closes the circuit of an electro-magnet, the armature of which operates the starter contacts. Thus a small current has the effect of switching the large current required by the motor. An additional benefit is that if the battery is low, it will not operate the solenoid, and will not be damaged by an excessive drain.

2 It is possible to tell if the solenoid is operating by listening for the 'click' as the starter button is pressed. If this is not to be heard, and the battery is fully charged, check the solenoid coil circuit for a fault. The solenoid coil itself may be checked for continuity across the yellow with red, and black leads to the solenoid, after disconnecting them.

3 If a click is heard, but the starter does not operate, check the security of the starter cable terminals. After long use the heavy duty switch contacts will become pitted or burned and no longer conduct. Disconnect the starter cables, and check for continuity across the switch when the starter button is depressed. There is no way of repairing the solenoid switch.

4 The solenoid switch is located under the left-hand frame cover and is mounted in a rubber clip. Remove the switch after disconnecting both starter cables. The terminals are beneath the rubber cover.

19 Starter and emergency switches: examination

1 If the starter solenoid does not operate when the starter button is depressed, first check the obvious - that the emergency switch is in the 'ON' position. Next check the state of charge of the battery. If the lights are dim when switched on, the battery requires charging. On the 'F' models, the clutch cutout circuit should also be checked.

2 The emergency switch may be checked for continuity when 'ON', between the black and the black with white leads. These are in the connector cover under the tank of the CB 400F and CB 550F models, inside the headlamp shell on the CB 550 model. Disconnect the snap connectors before testing. The starter button may be tested for continuity when depressed; between yellow with red lead and earth on the CB 550 models and yellow with red and black on the others. The connectors are in the same place as for the emergency switch.

20 Clutch cutout circuit: testing

1 On the 'F' models, a clutch operated cutout in conjunction with a diode and the neutral indicator switch, prevents operation of the starter with the engine in gear and the clutch engaged. The diode forms part of a logic circuit, and is plugged into a socket under the frame cover on the left-hand side.

2 If the cutout circuit is suspected, its components may be checked individually. Check for continuity between the green and the green with red leads inside the headlamp shell, when the clutch is disengaged only. Unplug the silicon diode, and check for continuity only in the direction shown by the arrow on the case (ie; positive lead to the 'blunt' end). If there is either continuity or no continuity in **both** directions, the diode is faulty. Do not use the 10 to the power of 6 ohms range for this test or the diode may be damaged. Checking the neutral indicator switch is dealt with in Section 12 of this Chapter.

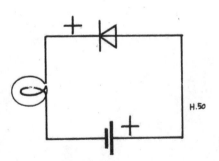

Fig. 6.4. Checking diode continuity. Lamp should light only when connected as shown

21.2 Fuse holder on 'F' models has snap-on cover

22.4a Lift on side next to spring to free bulb holder

22.4b Bulb is bayonet fitting in holder

22.5 The pilot bulb and holder are both bayonet fixing

23.2 Remove lens to renew stop/tail lamp

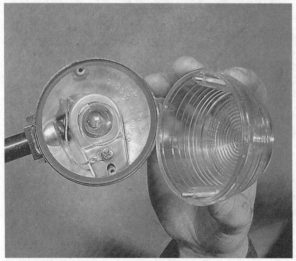

24.2 Both front and rear flashing indicators have same construction

25.2 Instrument lamp holders push into the instrument head

21 Fuses: replacing

1 A fuse is a weak link in the electrical circuit, which protects the system from damage by failing before the equipment or wires are overloaded. Consequently, a fuse should always be replaced with another of the rating specified by the manufacturer. **Never** should a fuse be replaced with wire, nails or so on. Do not replace a fuse before checking to discover the cause of failure of the original.
2 Before replacing a fuse, switch off the main switch. The fuses are located in a holder under the left-hand side cover. The CB 400F and CB 550F models have three fuses under a black plastic cover to the rear of the starter solenoid. The cover has one locating rag at one end and two the other. On the CB 550 model the fuse holder is cylindrical and held in a clip at the front of the compartment.
3 The CB 400F and CB 550F models have two 7 amp fuses protecting the headlamp and tail lamp circuits respectively and one 15 amp fuse protecting the remainder. A spare fuse of each rating is carried in the fuse holder. All are well marked. There is only one 15 amp fuse on the CB 550 model protecting the whole system. The spares are in the tool tray.
4 A blown fuse can be recognised by the melted metal strip, or a blackened appearance inside the glass tube. If a fuse blows, check and eliminate the fault. **Do not** put in a fuse of higher rating - another part may be damaged, or a fire result. The same applies to replacing the fuse with wire or silver paper, a 'get-you home' technique that should be used only in dire emergency. Provided the fault that caused the original fuse to blow has been eliminated a fuse 'doctored' in this fashion must be renewed with one of the correct rating at the earliest possible opportunity. Always carry spare fuses.

22 Headlamp: replacing bulbs and adjusting beam height

1 Only UK models have renewable lamp bulbs. The USA models have a sealed beam unit.
2 Remove the headlamp unit by unscrewing the two or three cross head screws around the headlamp shell. Remove the screws with spring washers and top hat bushes. Pull the rim off of the bottom of the shell, and unhook at the top.
3 Pull the three pin socket from the back of the sealed beam unit (USA models). To renew the unit, remove the beam adjusting screw and upper and lower retaining spring cotter pin and screw from the clamping ring.
   On some UK models with headlamps of the pre-focus type, the unit is held in place in the rim with spring clips. There is a locating tab to position the unit correctly. The pre-focus unit will only need removing if damage has occured. The units are marked 'TOP'.
4 To renew bulbs, pull the bulb holder upwards on the side by the spring and pull towards the spring to disengage a tongue from a slot in the reflector. Do not touch the glass envelope with the fingers ; use a cloth or tissue. The bulb is bayonet fitting with a push and turn. The bayonet tags are of different width for correct location. Make sure the contacts are clean.
5 The pilot light, fitted to UK models only, is bayonet fitting in the bulb holder, which in turn is bayonet fitting to the reflector.
6 **Do not** attempt to clean the reflector surface, or touch it with the fingers.
7 On models with a beam height adjusting screw on the left side of the rim (looking at the front), turn clockwise to raise the beam and vice versa. On other models, slacken the long hexagonal nuts on each side of the headlamp shell and pivot the shell.
8 UK lighting regulations stipulate that the lighting system must be arranged so that the light will not dazzle a person standing at a distance greater than 25 feet from the lamp, whose eye level is not less than 3 feet 6 inches above that plane. It is easy to approximate this setting by placing the machine 25 feet away from a wall, on a level road and setting the beam height so

that it is concentrated at the same height as the distance of the centre of the headlamp from the ground. The rider must be seated normally during this operation and also the pillion passenger, if one is carried regularly.

23 Stop and tail lamp: replacing bulbs

1 The rear lamp has a twin filament bulb. The stop lamp operates when both front and rear brakes are applied. In the UK, a working stop light is a statutory requirement.
2 To renew the bulb, first remove the lens after unscrewing the cross-head screws. Check the rubber gasket underneath the lens. The bulb is bayonet fitting, with offset pins so that it can be replaced in one position only. Do not touch the glass envelope; use a cloth or tissue. Make sure that the contacts are clean.
3 When replacing the lens, do not over-tighten the screws and crack the plastic.

24 Direction indicators: replacing bulbs

1 The front and rear indicators are of the same construction. Differences occur on the CB 400F and CB 550F models supplied to the USA market, where the front indicators are switched on when main beam is selected. This is achieved by using a dual filament lamp, as fitted to the rear light.
2 To renew the bulb, unscrew the two cross-head screws retaining the lens. Remove the lens, and check the rubber gasket. The bulb is bayonet fixing; the twin filament type has offset pins to ensure correct connection. Always use bulbs of the correct wattage, or the flashing rate will be affected. Do not touch the glass envelope; use a cloth or tissue. Make sure that the contacts are clean.
3 Replace the lens with its gasket. Do not overtighten the fixing screws and break the plastic.

25 Instrument heads: replacing bulbs

1 Each instrument has two illuminating bulbs. To renew these, the instrument has to be removed.
2 Unscrew the knurled drive cable nut and withdraw the cable. Remove the acorn nuts with large washers and rubber bushes and sleeves. Lift the instrument head and pull the bulb holders out of the base of the instrument

26.1 Warning lamps are bayonet fitting ('F' models illustrated)

3   Do not overtighten the cover fixing screws and strip the threads.

26 Warning lamps: CB 550 model only

1   The four warning lamps on the CB 550 model are mounted in a socket beneath the handlebar clamp.
2   Remove the four bolts and the handlebar clamp. Pull the bulb holder out of the bottom of the socket. The bulbs are bayonet fitting. Make sure the contacts are clean.
3   The bulbs are bayonet fitting and will press and turn. Clean the contacts.

27 Warning lamp console: CB 400F and CB 550F models

1   The five bulbs in this console may be changed when the lens cover is removed. This is retained by the three cross-head self-tapping screws in the top flange of the cover. Unscrew the screws and lift the cover.
2   The bulbs are bayonet fixing and quite difficult to remove, being a tight fit in the rubber surround. When replacing the bulbs, ensure that they are fully home.

28 Direction indicator relay: examination and renewal

1   A fault in the indicator circuit will be apparent from the incorrect flashing rate. If both lamps are of the correct wattage, check all wires and connections, particularly the earth connections. Do not forget the warning lamp or lamps.
2   If the relay malfunctions, the usual indication is one flash before the system goes dead. The relay should be heard clicking when working correct, it will need to be renewed if the fault cannot be traced elsewhere. Handle the relay with care, it may be damaged if dropped. Switch off the main switch before removing the relay.
3   Two types of relay are fitted, cylindrical or rectangular. Both are mounted in rubber, under the frame cover on the left-hand side. Pull off the leads after switching off, and remove the relay from its rubber clip. Note where the leads came from; there are coloured marks on the flasher can (black, grey and green), and replace them on the same terminals. The flashing rate in the UK must be between 65 and 90 flashes per minute. If the rate deviates, check bulb ratings, contacts and earths.
4   On the CB 550F model an audible indicator warning is also fitted. This is mounted on the left-hand fork cover. It cannot be serviced. Test as for the horn.

29 Fault diagnosis: electrical system

Symptom	Cause	Remedy
Complete electrical failure	Blown fuse(s)	Check wiring and electrical components for short circuit before fitting new fuse.
	Isolated battery	Check battery connections, also whether connections show signs of corrosion.
Dim lights, horn and starter inoperative	Discharged battery	Remove battery and charge with battery charger. Check generator output and voltage regulator settings.
Constantly blowing bulbs	Vibration or poor earth connection	Check security of bulb holders. Check earth return connections.
Parking lights dim rapidly	Battery will not hold charge	Renew battery at earliest opportunity.
Tail lamp fails (CB400F and 550F)	Blown bulb or fuse	Renew.
Headlamp fails (CB400F and 550F)	Blown bulb or fuse	Renew.
Flashing indicators do not operate, or flash fast or slow	Blown bulb Damaged flasher unit	Renew bulb. Renew flasher unit.
Horn inoperative or weak	Faulty switch Out of adjustment	Check switch. Re-adjust.
Incorrect charging	Faulty coil Faulty rectifier Faulty regulator Wiring fault	Check. Check. Check and adjust. Check.
Over or under-charging	As above, or faulty battery	Check.
Starter motor sluggish	Worn brushes Dirty commutator	Remove starter motor and renew brushes. Clean.
Starter motor does not turn	Machine in gear ('F' models only) Emergency switch in OFF position Faulty switches or wiring Battery flat	Disengage clutch. Turn on. Check continuity. Charge.

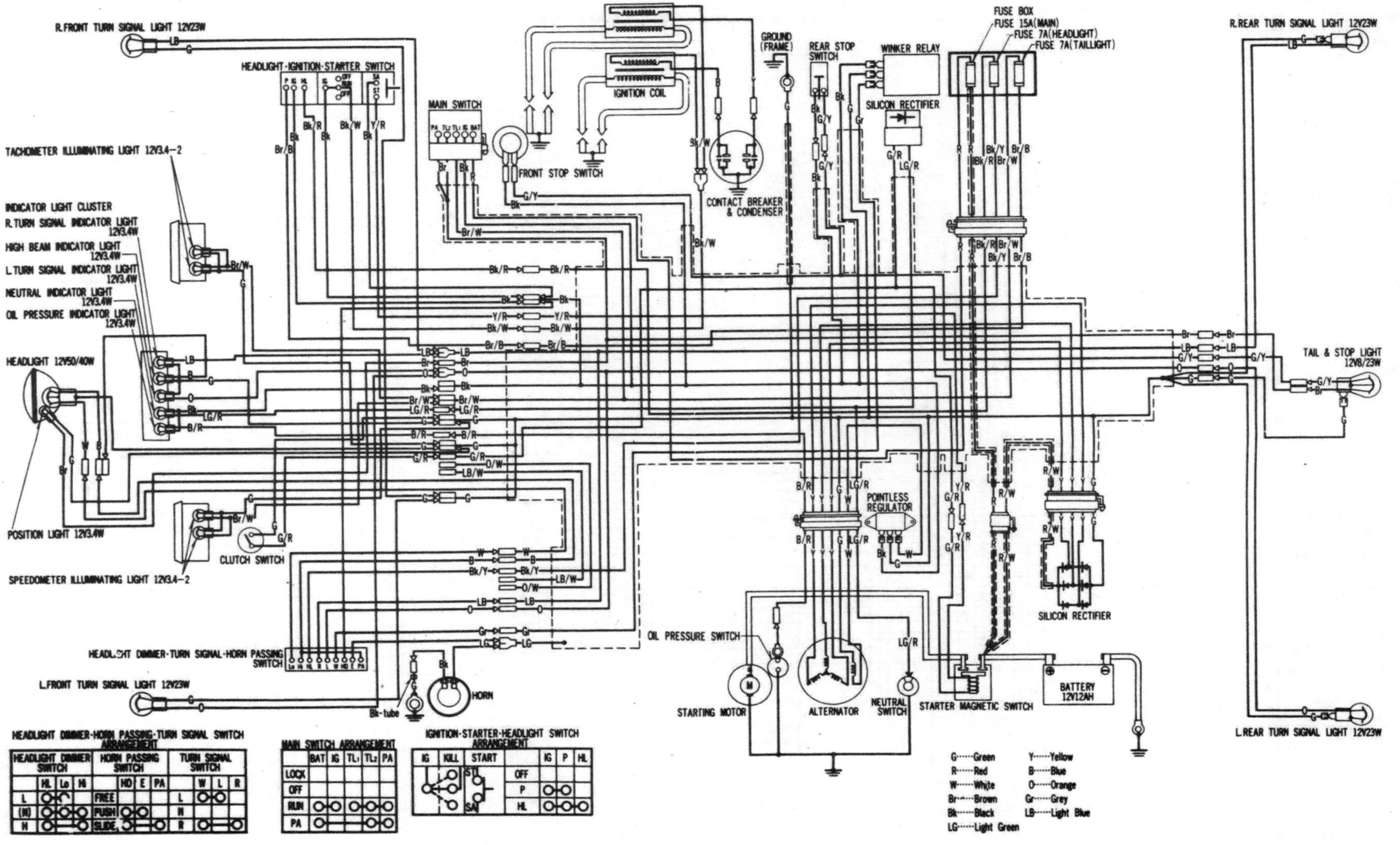

R.REAR TURN SIGNAL LIGHT 12V23W

TAIL & STOP LIGHT 12V8/23W

L.REAR TURN SIGNAL LIGHT 12V23W

SILICON RECTIFIER

BATTERY 12V12AH

STARTER MAGNETIC SWITCH

FUSE BOX
FUSE 15A (MAIN)
FUSE 7A (HEADLIGHT)
FUSE 7A (TAILLIGHT)

Y—Yellow
B—Blue
O—Orange
Gr—Grey
LB—Light Blue

G—Green
R—Red
W—White
Br—Brown
Bk—Black
LG—Light Green

WINKER RELAY

SILICON RECTIFIER

PONTLESS REGULATOR

NEUTRAL SWITCH

REAR STOP SWITCH

ALTERNATOR

OIL PRESSURE SWITCH

STARTING MOTOR

GROUND (FRAME)

CONTACT BREAKER & CONDENSER

IGNITION COIL

FRONT STOP SWITCH

MAIN SWITCH

HEADLIGHT-IGNITION-STARTER SWITCH

HORN

CLUTCH SWITCH

IGNITION-STARTER-HEADLIGHT SWITCH ARRANGEMENT

	IG	P	HL
OFF			
P			
HL			

	IG	KILL	START
ST			

MAIN SWITCH ARRANGEMENT

	BAT	IG	TLi	TLi	PA
LOCK					
OFF					
RUN					
PA					

HEADLIGHT DIMMER-HORN PASSING-TURN SIGNAL SWITCH

HEADLIGHT DIMMER-HORN PASSING-TURN SIGNAL-HORN SWITCH ARRANGEMENT

TURN SIGNAL SWITCH

			W	L	R
		L			
		N			
		R			

TACHOMETER ILLUMINATING LIGHT 12V3.4W

INDICATOR LIGHT CLUSTER
R.TURN SIGNAL INDICATOR LIGHT 12V3.4W
HIGH BEAM INDICATOR LIGHT 12V3.4W
L.TURN SIGNAL INDICATOR LIGHT 12V3.4W
NEUTRAL INDICATOR LIGHT 12V3.4W
OIL PRESSURE INDICATOR LIGHT 12V3.4W

R.FRONT TURN SIGNAL LIGHT 12V23W

HEADLIGHT 12V50/40W

POSITION LIGHT 12V4W

SPEEDOMETER ILLUMINATING LIGHT 12V3.4W

L.FRONT TURN SIGNAL LIGHT 12V23W

Fig. 6.5. Wiring diagram for CB 400F (UK models)

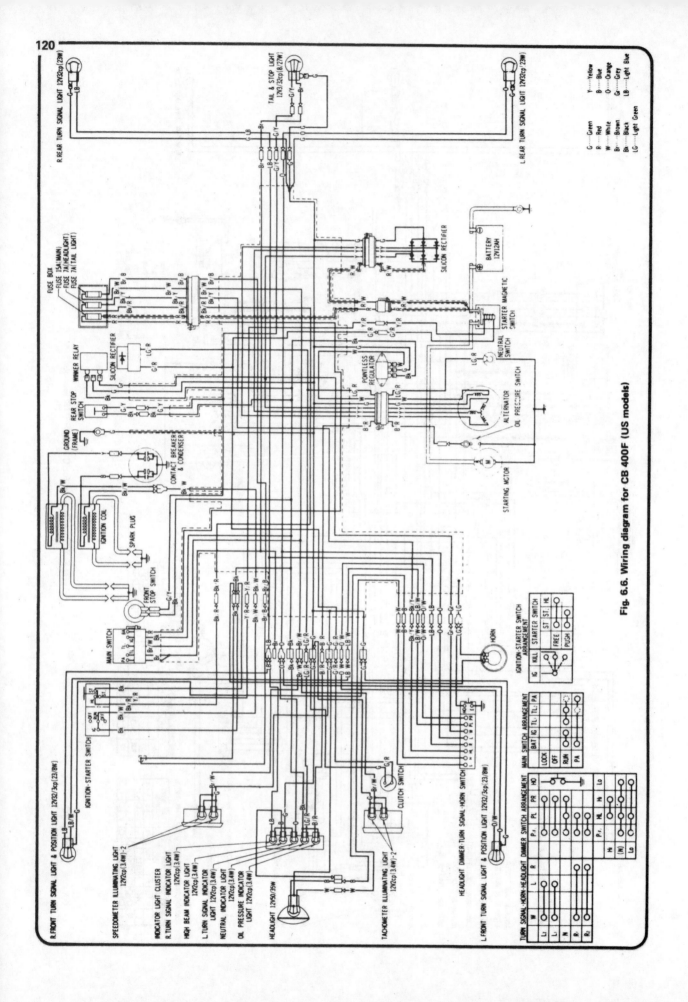

Fig. 6.6. Wiring diagram for CB 400F (US models)

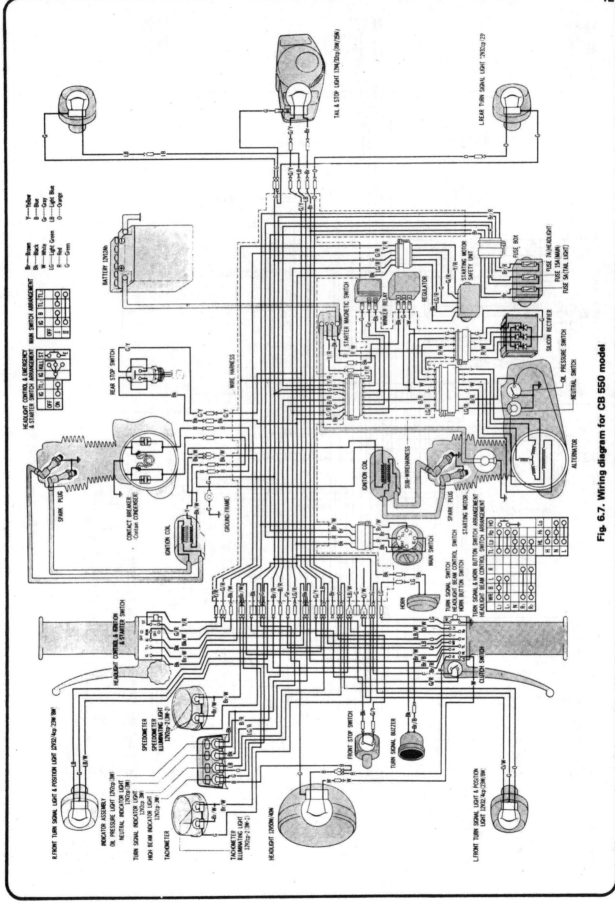

Fig. 6.7. Wiring diagram for CB 550 model

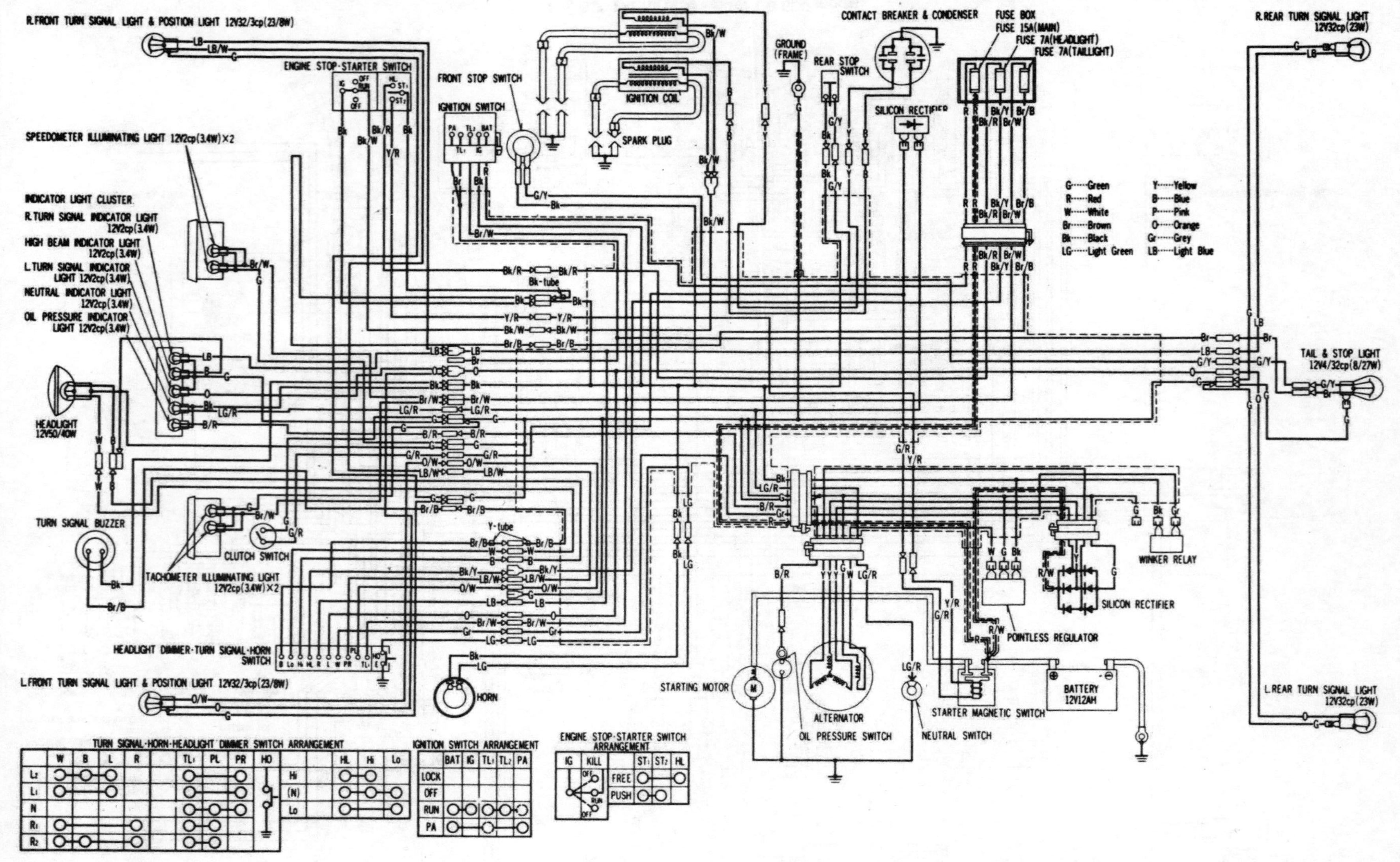

Fig. 6.8. Wiring diagram for CB 550F model

Metric conversion tables

Inches	Decimals	Millimetres	Millimetres to Inches		Inches to Millimetres	
			mm	Inches	Inches	mm
1/64	0.015625	0.3969	0.01	0.00039	0.001	0.0254
1/32	0.03125	0.7937	0.02	0.00079	0.002	0.0508
3/64	0.046875	1.1906	0.03	0.00118	0.003	0.0762
1/16	0.0625	1.5875	0.04	0.00157	0.004	0.1016
5/64	0.078125	1.9844	0.05	0.00197	0.005	0.1270
3/32	0.09375	2.3812	0.06	0.00236	0.006	0.1524
7/64	0.109375	2.7781	0.07	0.00276	0.007	0.1778
1/8	0.125	3.1750	0.08	0.00315	0.008	0.2032
9/64	0.140625	3.5719	0.09	0.00354	0.009	0.2286
5/32	0.15625	3.9687	0.1	0.00394	0.01	0.254
11/64	0.171875	4.3656	0.2	0.00787	0.02	0.508
3/16	0.1875	4.7625	0.3	0.01181	0.03	0.762
13/64	0.203125	5.1594	0.4	0.01575	0.04	1.016
7/32	0.21875	5.5562	0.5	0.01969	0.05	1.270
15/64	0.234375	5.9531	0.6	0.02362	0.06	1.524
1/4	0.25	6.3500	0.7	0.02756	0.07	1.778
17/64	0.265625	6.7469	0.8	0.03150	0.08	2.032
9/32	0.28125	7.1437	0.9	0.03543	0.09	2.286
19/64	0.296875	7.5406	1	0.03947	0.1	2.54
5/16	0.3125	7.9375	2	0.07874	0.2	5.08
21/64	0.328125	8.3344	3	0.11811	0.3	7.62
11/32	0.34375	8.7312	4	0.15748	0.4	10.16
23/64	0.359375	9.1281	5	0.19685	0.5	12.70
3/8	0.375	9.5250	6	0.23622	0.6	15.24
25/64	0.390625	9.9219	7	0.27559	0.7	17.78
13/32	0.40625	10.3187	8	0.31496	0.8	20.32
27/64	0.421875	10.7156	9	0.35433	0.9	22.86
7/16	0.4375	11.1125	10	0.39370	1	25.4
29/64	0.453125	11.5094	11	0.43307	2	50.8
15/32	0.46875	11.9062	12	0.47244	3	76.2
31/64	0.484375	12.3031	13	0.51181	4	101.6
1/2	0.5	12.7000	14	0.55118	5	127.0
33/64	0.515625	13.0969	15	0.59055	6	152.4
17/32	0.53125	13.4937	16	0.62992	7	177.8
35/64	0.546875	13.8906	17	0.66929	8	203.2
9/16	0.5625	14.2875	18	0.70866	9	228.6
37/64	0.578125	14.6844	19	0.74803	10	254.0
19/32	0.59375	15.0812	20	0.78740	11	279.4
39/64	0.609375	15.4781	21	0.82677	12	304.8
5/8	0.625	15.8750	22	0.86614	13	330.2
41/64	0.640625	16.2719	23	0.90551	14	355.6
21/32	0.65625	16.6687	24	0.94488	15	381.0
43/64	0.671875	17.0656	25	0.98425	16	406.4
11/16	0.6875	17.4625	26	1.02362	17	431.8
45/64	0.703125	17.8594	27	1.06299	18	457.2
23/32	0.71875	18.2562	28	1.10236	19	482.6
47/64	0.734375	18.6531	29	1.14173	20	508.0
3/4	0.75	19.0500	30	1.18110	21	533.4
49/64	0.765625	19.4469	31	1.22047	22	558.8
25/32	0.78125	19.8437	32	1.25984	23	584.2
51/64	0.796875	20.2406	33	1.29921	24	609.6
13/16	0.8125	20.6375	34	1.33858	25	635.0
53/64	0.828125	21.0344	35	1.37795	26	660.4
27/32	0.84375	21.4312	36	1.41732	27	685.8
55/64	0.859375	21.8281	37	1.4567	28	711.2
7/8	0.875	22.2250	38	1.4961	29	736.6
57/64	0.890625	22.6219	39	1.5354	30	762.0
29/32	0.90625	23.0187	40	1.5748	31	787.4
59/64	0.921875	23.4156	41	1.6142	32	812.8
15/16	0.9375	23.8125	42	1.6535	33	838.2
61/64	0.953125	24.2094	43	1.6929	34	863.6
31/32	0.96875	24.6062	44	1.7323	35	889.0
63/64	0.984375	25.0031	45	1.7717	36	914.4

English/American terminology

Because this book has been written in England, British English component names, phrases and spellings have been used throughout. American English usage is quite often different and whereas normally no confusion should occur, a list of equivalent terminology is given below.

English	American	English	American
Air filter	Air cleaner	Number plate	License plate
Alignment (headlamp)	Aim	Output or layshaft	Countershaft
Allen screw/key	Socket screw/wrench	Panniers	Side cases
Anticlockwise	Counterclockwise	Paraffin	Kerosene
Bottom/top gear	Low/high gear	Petrol	Gasoline
Bottom/top yoke	Bottom/top triple clamp	Petrol/fuel tank	Gas tank
Bush	Bushing	Pinking	Pinging
Carburettor	Carburetor	Rear suspension unit	Rear shock absorber
Catch	Latch	Rocker cover	Valve cover
Circlip	Snap ring	Selector	Shifter
Clutch drum	Clutch housing	Self-locking pliers	Vise-grips
Dip switch	Dimmer switch	Side or parking lamp	Parking or auxiliary light
Disulphide	Disulfide	Side or prop stand	Kick stand
Dynamo	DC generator	Silencer	Muffler
Earth	Ground	Spanner	Wrench
End float	End play	Split pin	Cotter pin
Engineer's blue	Machinist's dye	Stanchion	Tube
Exhaust pipe	Header	Sulphuric	Sulfuric
Fault diagnosis	Trouble shooting	Sump	Oil pan
Float chamber	Float bowl	Swinging arm	Swingarm
Footrest	Footpeg	Tab washer	Lock washer
Fuel/petrol tap	Petcock	Top box	Trunk
Gaiter	Boot	Torch	Flashlight
Gearbox	Transmission	Two/four stroke	Two/four cycle
Gearchange	Shift	Tyre	Tire
Gudgeon pin	Wrist/piston pin	Valve collar	Valve retainer
Indicator	Turn signal	Valve collets	Valve cotters
Inlet	Intake	Vice	Vise
Input shaft or mainshaft	Mainshaft	Wheel spindle	Axle
Kickstart	Kickstarter	White spirit	Stoddard solvent
Lower leg	Slider	Windscreen	Windshield
Mudguard	Fender		

Safety first!

Professional motor mechanics are trained in safe working procedures. However enthusiastic you may be about getting on with the job in hand, do take the time to ensure that your safety is not put at risk. A moment's lack of attention can result in an accident, as can failure to observe certain elementary precautions.

There will always be new ways of having accidents, and the following points do not pretend to be a comprehensive list of all dangers; they are intended rather to make you aware of the risks and to encourage a safety-conscious approach to all work you carry out on your vehicle.

Essential DOs and DON'Ts

DON'T start the engine without first ascertaining that the transmission is in neutral.

DON'T suddenly remove the filler cap from a hot cooling system – cover it with a cloth and release the pressure gradually first, or you may get scalded by escaping coolant.

DON'T attempt to drain oil until you are sure it has cooled sufficiently to avoid scalding you.

DON'T grasp any part of the engine, exhaust or silencer without first ascertaining that it is sufficiently cool to avoid burning you.

DON'T allow brake fluid or antifreeze to contact the machine's paintwork or plastic components.

DON'T syphon toxic liquids such as fuel, brake fluid or antifreeze by mouth, or allow them to remain on your skin.

DON'T inhale dust – it may be injurious to health (see *Asbestos* heading).

DON'T allow any spilt oil or grease to remain on the floor – wipe it up straight away, before someone slips on it.

DON'T use ill-fitting spanners or other tools which may slip and cause injury.

DON'T attempt to lift a heavy component which may be beyond your capability – get assistance.

DON'T rush to finish a job, or take unverified short cuts.

DON'T allow children or animals in or around an unattended vehicle.

DON'T inflate a tyre to a pressure above the recommended maximum. Apart from overstressing the carcase and wheel rim, in extreme cases the tyre may blow off forcibly.

DO ensure that the machine is supported securely at all times. This is especially important when the machine is blocked up to aid wheel or fork removal.

DO take care when attempting to slacken a stubborn nut or bolt. It is generally better to pull on a spanner, rather than push, so that if slippage occurs you fall away from the machine rather than on to it.

DO wear eye protection when using power tools such as drill, sander, bench grinder etc.

DO use a barrier cream on your hands prior to undertaking dirty jobs – it will protect your skin from infection as well as making the dirt easier to remove afterwards; but make sure your hands aren't left slippery. Note that long-term contact with used engine oil can be a health hazard.

DO keep loose clothing (cuffs, tie etc) and long hair well out of the way of moving mechanical parts.

DO remove rings, wristwatch etc, before working on the vehicle – especially the electrical system.

DO keep your work area tidy – it is only too easy to fall over articles left lying around.

DO exercise caution when compressing springs for removal or installation. Ensure that the tension is applied and released in a controlled manner, using suitable tools which preclude the possibility of the spring escaping violently.

DO ensure that any lifting tackle used has a safe working load rating adequate for the job.

DO get someone to check periodically that all is well, when working alone on the vehicle.

DO carry out work in a logical sequence and check that everything is correctly assembled and tightened afterwards.

DO remember that your vehicle's safety affects that of yourself and others. If in doubt on any point, get specialist advice.

IF, in spite of following these precautions, you are unfortunate enough to injure yourself, seek medical attention as soon as possible.

Asbestos

Certain friction, insulating, sealing, and other products – such as brake linings, clutch linings, gaskets, etc – contain asbestos. *Extreme care must be taken to avoid inhalation of dust from such products since it is hazardous to health.* If in doubt, assume that they *do* contain asbestos.

Fire

Remember at all times that petrol (gasoline) is highly flammable. Never smoke, or have any kind of naked flame around, when working on the vehicle. But the risk does not end there – a spark caused by an electrical short-circuit, by two metal surfaces contacting each other, by careless use of tools, or even by static electricity built up in your body under certain conditions, can ignite petrol vapour, which in a confined space is highly explosive.

Always disconnect the battery earth (ground) terminal before working on any part of the fuel or electrical system, and never risk spilling fuel on to a hot engine or exhaust.

It is recommended that a fire extinguisher of a type suitable for fuel and electrical fires is kept handy in the garage or workplace at all times. Never try to extinguish a fuel or electrical fire with water.

Note: *Any reference to a 'torch' appearing in this manual should always be taken to mean a hand-held battery-operated electric lamp or flashlight. It does **not** mean a welding/gas torch or blowlamp.*

Fumes

Certain fumes are highly toxic and can quickly cause unconsciousness and even death if inhaled to any extent. Petrol (gasoline) vapour comes into this category, as do the vapours from certain solvents such as trichloroethylene. Any draining or pouring of such volatile fluids should be done in a well ventilated area.

When using cleaning fluids and solvents, read the instructions carefully. Never use materials from unmarked containers – they may give off poisonous vapours.

Never run the engine of a motor vehicle in an enclosed space such as a garage. Exhaust fumes contain carbon monoxide which is extremely poisonous; if you need to run the engine, always do so in the open air or at least have the rear of the vehicle outside the workplace.

The battery

Never cause a spark, or allow a naked light, near the vehicle's battery. It will normally be giving off a certain amount of hydrogen gas, which is highly explosive.

Always disconnect the battery earth (ground) terminal before working on the fuel or electrical systems.

If possible, loosen the filler plugs or cover when charging the battery from an external source. Do not charge at an excessive rate or the battery may burst.

Take care when topping up and when carrying the battery. The acid electrolyte, even when diluted, is very corrosive and should not be allowed to contact the eyes or skin.

If you ever need to prepare electrolyte yourself, always add the acid slowly to the water, and never the other way round. Protect against splashes by wearing rubber gloves and goggles.

Mains electricity and electrical equipment

When using an electric power tool, inspection light etc, always ensure that the appliance is correctly connected to its plug and that, where necessary, it is properly earthed (grounded). Do not use such appliances in damp conditions and, again, beware of creating a spark or applying excessive heat in the vicinity of fuel or fuel vapour. Also ensure that the appliances meet the relevant national safety standards.

Ignition HT voltage

A severe electric shock can result from touching certain parts of the ignition system, such as the HT leads, when the engine is running or being cranked, particularly if components are damp or the insulation is defective. Where an electronic ignition system is fitted, the HT voltage is much higher and could prove fatal.

Index